T0137090

Lecture Notes in Social Networks

Series Editors

Reda Alhajj, University of Calgary, Calgary, AB, Canada
Uwe Glässer, Simon Fraser University, Burnaby, BC, Canada

Advisory Board

Charu C. Aggarwal, IBM T.J. Watson Research Center, Hawthorne, NY, USA
Patricia L. Brantingham, Simon Fraser University, Burnaby, BC, Canada
Thilo Gross, University of Bristol, Bristol, UK
Jiawei Han, University of Illinois at Urbana-Champaign, IL, USA
Huan Liu, Arizona State University, Tempe, AZ, USA
Raul Manasevich, University of Chile, Santiago, Chile
Anthony J. Masys, Centre for Security Science, Ottawa, ON, Canada
Carlo Morselli, University of Montreal, QC, Canada
Rafael Wittek, University of Groningen, The Netherlands
Daniel Zeng, The University of Arizona, Tucson, AZ, USA

More information about this series at http://www.springer.com/series/8768

Jalal Kawash • Nitin Agarwal • Tansel Özyer
Editors

Prediction and Inference from Social Networks and Social Media

 Springer

Editors
Jalal Kawash
Department of Computer Science
University of Calgary
Calgary, AB, Canada

Nitin Agarwal
Information Science Department
University of Arkansas at Little Rock
Little Rock, AR, USA

Tansel Özyer
Department of Computer Engineering
TOBB University
Ankara, Turkey

ISSN 2190-5428   ISSN 2190-5436   (electronic)
Lecture Notes in Social Networks
ISBN 978-3-319-84553-1   ISBN 978-3-319-51049-1   (eBook)
DOI 10.1007/978-3-319-51049-1

© Springer International Publishing AG 2017
Softcover reprint of the hardcover 1st edition 2017
This work is subject to copyright. All rights are reserved by the Publisher, whether the whole or part of the material is concerned, specifically the rights of translation, reprinting, reuse of illustrations, recitation, broadcasting, reproduction on microfilms or in any other physical way, and transmission or information storage and retrieval, electronic adaptation, computer software, or by similar or dissimilar methodology now known or hereafter developed.
The use of general descriptive names, registered names, trademarks, service marks, etc. in this publication does not imply, even in the absence of a specific statement, that such names are exempt from the relevant protective laws and regulations and therefore free for general use.
The publisher, the authors and the editors are safe to assume that the advice and information in this book are believed to be true and accurate at the date of publication. Neither the publisher nor the authors or the editors give a warranty, express or implied, with respect to the material contained herein or for any errors or omissions that may have been made.

Printed on acid-free paper

This Springer imprint is published by Springer Nature
The registered company is Springer International Publishing AG
The registered company address is: Gewerbestrasse 11, 6330 Cham, Switzerland

Preface

Social networks (SN) have brought an unprecedented revolution in how people interact and socialize. SN are used not only as a lifestyle but also in various other domains, including medicine, business, education, politics, and activism. SN have as well grown in sizes to include billions of users. As of this writing, Twitter claims to have 313 million monthly active users. Facebook grew by the end of 2016 to 1.71 billion users with 1.13 billion daily active users. Facebook now has a "population" that surpassed the population of India! Online social media (OSM), media produced by SN users, have offered a real and viable alternative to conventional mainstream media. OSM are likely to provide "raw", unedited information, and the details can be overwhelming with the potential of misinformation and disinformation. Yet, OSM are leading to the democratization of knowledge and information. OSM are allowing almost any citizen to become a journalist reporting on specific events of interest. This is resulting in unimaginable amounts of information being shared among a huge number of OSM participants. For example, Facebook users are generating several billion "likes" and several hundred million posted pictures in a single day. Twitter users are producing 6000 tweets per second. This immense amount of OSM poses increasing challenges to mine, analyse, utilize, and exploit such content. One grand challenge in OSM is mining its content to make useful inferences or predict future behaviour of SN users. This book includes nine contributions that examine new approaches that relate to predication and inference in OSM content. What follows is a quick summary of these chapters included in this book.

Mood prediction in SN is of great utility, especially in the medical field. For instance, predicting a patient's mood can be crucial to identify signs for depression. In this book, Roshanaei et al.'s approach is to design accurate personalized classifiers that can predict a person's emotions based on features extracted from OSM postings. By developing techniques to mine features such as social engagement, gender, language and linguistic styles, and various psychological features in a patient's tweets, they are able to infer the patient's mode as positive, negative, or neutral. In a different chapter, Kaya investigates the prediction of future symptoms of patients from current patients' records. Kaya's approach consists of the construction of a weighted symptom network and, then, through unsupervised link prediction, building the evolving structure of symptom network with respect to patients' ages. Medical SN are also the subject of a chapter by Ayadi et al. Their objective is the automatic inference of indexing medical images. Their approach automatically extracts and analyses information from specialist's analysis and recommendations. The approach is multilingual, applying to different languages.

Shahriari et al. study signed SN in another chapter. Their focus is the significance of overlapping members in signed networks, and the intension is to discover overlapping communities in these networks. Different features are used to investigate the significance of overlapping members.

Alhajj looks at link prediction as a class of recommendation systems, predicting recommendations (links) between users and items. Efficiently finding hidden links and extracting missing information in a network aid in identifying a set of new interactions. In this chapter, Alhajj approaches the problem by exploiting the benefits of social network analysis tools and algorithms. For better scalability and efficiency, Alhajj utilizes a graph database model, as opposed to a traditional relational database.

The prediction of the quality of Wikipedia articles is studied in a chapter by De La Robertie et al. Wikipedia is not immune to problems relating to article quality, such as reputability of third-party sources and vandalism. The huge number of articles and the intensive edit rate make it infeasible to manually evaluate the content quality. De La Robertie et al. propose a quality model that integrates both temporal and structural features captured from the implicit peer review process enabled by Wikipedia. Two mutually reinforced factors are taken into account: article quality and author authority.

Charitonidis et al. realize that microblogs, such as Twitter, can be used for a good or a bad cause. In this chapter, their concern is the prediction of collective behaviour before it happens. Their approach is to analyse social media content to detect what they call "weak" signals; these are indicators that initially appear isolated but can be early indicators of large-scale, real-world phenomena. The 2011 London riots and tweets pertaining to them are their test-bed for their study.

Discussion forums of Massive Open Online Courses (MOOCs) are the subject of a chapter by Hecking et al. Their objective is to infer the structure of knowledge exchange in MOOC forums. The first step is the extraction of dynamic communication networks from forum data. Next is characterizing users according to information-seeking and information-giving behaviours and analysing the development of individual actors. Finally comes the reduction of a dynamic network to reflect knowledge exchange between clusters of actors and its evaluation.

Bouanan argues that the definition of a unique social network is too restrictive since in reality people are interlinked by several relationships, rather than by only one relationship. Hence, multidimensional networks are a better representation of relations among humans. They also define distinctive rules for the simulation of message diffusion. Hence, Bouanan's model includes agents interacting through multiple channels or with different relationships, and information disseminates differently on different link categories. The modelling and simulation of multidimensional networks are the subject of this chapter.

Calgary, AB, Canada   Jalal Kawash
Little Rock, AR, USA   Nitin Agarwal
Ankara, Turkley   Tansel Özyer

Contents

Chapter 1
Having Fun?: Personalized Activity-Based Mood Prediction in Social Media

Mahnaz Roshanaei, Richard Han, and Shivakant Mishra

1 Introduction

Positivity and negativity attributes of a person's mood and emotions are reflected in his or her interactions in his/her daily life. The question here is how much a person's mood and emotions are effected by their personal activates? How much these activities are personalized? For example, if working effects on person's mood as happy or sad.

On the other hand, usage of social networks has exploded over the past decade or so. Users now routinely share their thought, opinions, feelings as well as their daily activities on various social networks. In fact, there is evidence that even people leading otherwise a secluded life do indulge in online social activities. An interesting consequence of this explosive usage of social networks is that it is possible to glean the current mood and emotion of a user from his or her social network postings. A question that arises in this context is: Can we use any differentiating features exhibited by people on their online social activities to build appropriate classifiers that can identify the positivity or negativity of users with high accuracy and low false positive and negative rates?

Negative emotions can have disastrous consequences leading to severe depressions, family neglect, violent behavior, criminal activities, and even suicides. As a result, it is important to identify people suffering from negative emotions in a timely manner, so that important help and support can be provided to them.

Recent work discovered visual, locative, temporal, and social context as four types of cues to trigger memories of events and associated emotions [1]. In [2], we provided a detailed analysis of positivity and negativity attributes of user postings on Twitter. Our study is based on an analysis of a Twitter dataset published by the

---

M. Roshanaei (✉) • R. Han • S. Mishra
Department of Computer Science, University of Colorado, Boulder, CO 80309-0430, USA
e-mail: mahnaz.roshanaei@colorado.edu; richard.han@colorado.edu; mishras@colorado.edu

© Springer International Publishing AG 2017
J. Kawash et al. (eds.), *Prediction and Inference from Social Networks and Social Media*, Lecture Notes in Social Networks, DOI 10.1007/978-3-319-51049-1_1

University of Illinois [3]. An important finding of this work was that social media contains useful behavioral cues to classify users into positive and negative groups based on network density and degree of social activity either in information sharing or emotional interaction and social awareness.

These findings inspired us to implement a personalized activity-based classifier which can assist individuals in predicting their mood and emotions from their social network postings over time. Personalized classification is important, since one person's fun hobby may be detested by another person drudgery, e.g., rock-climbing or dancing or shopping, etc. We also considered the effect of temporal nature of users positing as daily or weekly pattern on person's mood and emotion. Such a tool can improve people's emotional memory and their awareness of how daily activities and time may affect their emotion.

In the first step, we have developed a general classifier by using several typical features such as n-grams, emoticons, and abbreviation and word lengths. Then, we include the social network behavioral attributes of users such as the number of tweets, retweets, followers, and friends in the context of positive and negative users. We also analyze gender and psychological features again in the context of positive, negative, and neutral tweets. All of these features are used to build our general classifier and improve the overall accuracy. We show in this paper that training and testing a generalized classifier based on aggregating data from many users results in limited performance when using activity as an extra input feature. In the second step, we discuss the relationship between positive and negative attributes of users and their activities at the time of tweeting. We observe that each user has specific activities in correlation with his/her mood. We also find that the temporal nature of tweeting for each individual user is different. We use these observations to build a personalized classifier that identifies users' emotional states based on the history of their activities in addition to posting time.

The novelty/contributions of this paper are:

- This paper is the first to use activity as an input feature to predict mood,
- This paper is the first to show that it is important to design personalized classifiers rather than generalized classifiers to achieve individualized mood prediction, and
- This paper is the first to demonstrate that temporal information is also effective in predicting mood in personalized classifiers.

The rest of the paper is organized as follows. In Sect. 2, we discuss prior work related to the intersection of online social media and mood. In Sect. 3, we describe the Twitter dataset used in our analysis as well as how we have cleaned and labeled this dataset. Sect. 4 describes the features that are used to design our classifier. The methodology deployed in implementing the classifier and the related results for both general and personalized are presented in Sect. 5. Finally, in Sect. 6, we conclude the paper.

2 Related Work

Most of the studies related to emotion and informatics are in the Human Computer Interaction (HCI) (e.g., see [4]). Based on experiments, it has been shown that an emotional state has the influential effect on a person's behavior and his/her interactions with other group members [5–7].

There has also been research done in the field of computational linguistics to detect emotions in text. Sentiment analysis, which is the task of identifying positive and negative opinions, emotions, and evaluation, has attracted the attention of a large number of researchers in this area. There has recently been some work by researchers in the area of phrase level and sentence level sentiment [8]. Considering online posting as a daily user activity, social media can be used to understand the behavioral attributes of the individual users [9–11].

Twitter as a public social network recently attracted the researchers' attention. There are several studies for tweets sentiment analysis in the literature [12]. In addition to linguistic features, other features such as emoticons and hashtags can be used to identify tweet polarity. In addition, some approaches that integrate lexicons, part-of-speech, and writing style have been studied [13, 14]. Designing the most appropriate classifiers to provide the highest accuracy has been considered as a new challenge in sentiment analysis [12]. In these papers text-based features are defined to reach the highest value of accuracy. In addition to sentiment analysis, emoticons and mood can be used to describe users' current experience and feeling. These features can also be used to identify the popularity and influence of each individual user in social network. For instance, paper [11] shows emoticon has a powerful effect to predict the social status on both Twitter and Facebook. Culture, language, and weather have also been shown to influence status updates and can illustrate user's feelings and mood [15, 16].

Social media has been recently considered as a good resource for predicting various aspects of mental health [17]. These studies illustrate the importance of social media as a potential for healthcare monitoring system and a promising tool for public health.

A number of researchers have illustrated the use of social media to diagnose depression. For instance, reference [18] investigates how language usage can reveal depression conditions using LIWC (Linguistic Inquiry and Word Count), where LIWC is a text analysis program that counts words in psychologically meaningful categories. An interesting study recently identified patterns of Internet usage and analyzed the online logs of undergraduate students to indicate depression [19]. This study shows that depressed students are more interested in using file-sharing services, sending emails, and chatting online compared to the students without signs of depression.

In [20–22], authors have defined several behavioral attributes of depressed and nondepressed individuals. They defined the application of Twitter as an information consuming and sharing tool for nondepressed individuals, and the second one is a tool for social awareness and emotional interaction for depressed individuals.

In [20], several attributes such as emotion, language and linguistic styles, ego network, and mentions of antidepressant medications to build a statistical classifier are presented. These features are obtained considering social interaction of 476 users in Twitter over a year contains 171 users with the positive sign of depression and 305 users with the negative sign. All these features are used to design a general classifier to diagnose depression in individual users. These studies can be used to develop future tools to identify users with depression signs and help them provide the right type of social interactions in online social media.

There is also a recent research shows how visual, locative, temporal, and social contexts can trigger memories of events and associated emotions [1]. For example, authors in paper [23] show how location can be considered as a fundamental contextual trigger for emotions [23]. Their observations show that the regular daily activities will occur at particular locations. Following these researches, our system extracts user's activities form using LIWC from their tweets. It also shows the importance of temporal information and daily activities in predicting each individual's mood and emotions.

3   Social Media Data

3.1   Twitter Dataset

For our study, we have used the Twitter dataset published by the University of Illinois [3]. As we mentioned in our previous paper [2], this dataset originally contained 284 million following relationships, 3 million user profiles, and 50 million tweets. Next, 150,000 users that have up to 500 tweets are filtered [3]. After cleaning the duplicated tweets and filtering the user IDs that were not present in the dataset, either in following relationships or user profiles, the number of tweets and user profiles was reduced to 45,864,434 and 131,010, respectively [2].

The maximum number of tweets in this dataset has been posted by the users in the year of 2011, January to June. Since, one of our goals is to analyze correlation between mood and the temporal nature of user posting, we decided to select the user IDs with sufficiently large number of tweets in the year 2011. The average number of tweets in 2011 per user is 241.3. To have a better tweets' posting distribution, we come up with selecting the user IDs with at least 180 tweets (one tweet per day) in the first 6-month period of year 2011. This filtering reduced the number of user IDs to 441. Details of the final dataset are shown in Table 1.1.

Table 1.1 Dataset in 2011

Number of unique users	441
Number of tweets	116,466
Average number of tweets per user	241.30
Average number of friends in Twitter	2034.1
Average number of followers in Twitter	2293.3
Average number of friends in original dataset	19.52
Average number of followers in original dataset	21.87

3.2 Ground Truth

The first requirement for designing classifiers to predict attributes of tweets is to get ground truth data that we have collected. We need to categorize each tweet as positive, negative, or neutral. Since, deciding whether a tweet is positive, negative, or neutral depends on many different factors and is subjective to a large extent, we decided to seek help from general users. We employed crowdworkers recruited through Amazon's Mechanical Turk.

As we mentioned in the previous section, our final dataset includes 441 user IDs.

As we want to have a consistent analysis to explore temporal nature of posting, we filtered our dataset to the user IDs that just posted tweets in only the first 6 months of 2011, January to June. We also filtered some users that write a lot of tweets as account of newspapers, companies or bloggers, etc., this filtering reduced the number of our user IDs and tweets to 142 and 111,930, respectively. Then, we identified user IDs with at least 400 tweets in this first 6 month of 2011 which resulted in 36 users. From this dataset, we randomly selected 18 user IDs with a total of 7255 tweets. The crowdworkers labeled each of these tweets as positive, negative, or neutral. To ensure the validity of labeling done by the crowdworkers, we chose only those crowdworkers that had a minimum of 95 % approval rating, English language proficiency, and familiarity with using Twitter. In addition, to address the subjective nature of our labeling, we got labeling of each tweet by five different crowdworker. Furthermore, we considered a tweet as positive, negative, or neutral only when there was an agreement in the labeling by at least four workers, i.e., we considered a tweet positive only when at least four out of five workers labeled it as positive. Similarly, we considered a tweet negative (neutral) only when four out of five workers labeled it negative (neutral). There were some tweets for which there was no such agreement among the workers. For instance, some tweets were categorized as positive by two workers, neutral by two other workers, and negative by a fifth worker. To resolve the categorization of such tweets, we got additional labeling from two more crowdworkers, and decided on a final label for those tweets based on majority. In the end, 2222 tweets were labeled based on 100% agreement, 4538 tweets were labeled based on 80% agreement, and the rest of the tweets were labeled based on majority agreement. This final labeled dataset is used as the training set to design our classifier.

4   Features

In order to design our classifier, we considered several types of features. Some of these features are described below.

Psychological Features We used Linguistic Inquiry and Word Count, LIWC (http://www.liwc.net/) to determine psychological value of a tweet. This tool contains a dictionary of thousands of words in which each word could be classified in one of six different categories of social, affective, cognitive, perceptual, biological processes, and relativity [24]. LIWC has often been used for studies on variations in language use across different people. Published papers show that LIWC has been validated to perform well in studies on variations in language use across different people [21]. The psychological process in LIWC maps words onto 25 psychological dimensions.

Figure 1.1 shows the relationship between the 25dimensions of normalized psychological processes derived from LIWC and the three categories of positive, negative, and neutral tweets. We see that for some of these psychological dimensions listed on the x-axis, variation in the number of positive/negative/neutral tweets is quite pronounced. This observation can be used to improve the performance of our classifier design by reducing the number of psychological dimensions.

Diurnal and Weekly Activities In our analysis, we explored the diurnal pattern of posting of the three categories (positive, negative, and neutral) of tweets over a day as well as different days of a week. As we want to have a consistent analysis to explore temporal nature of posting, we used that 142 user IDs that just posted tweets in only the first 6 month of 2011, January to June.

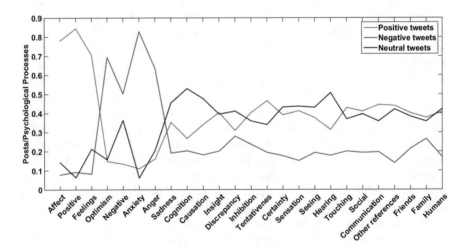

Fig. 1.1 Trends for the normalized psychological processes values corresponding to all classes

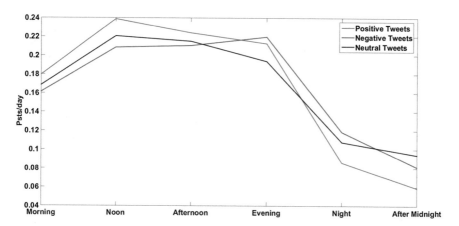

Fig. 1.2 Diurnal trends of posts corresponding to all classes

Table 1.2 Time interval per day

Hours	Interval
6:00–10:00 AM	Morning
10:00 AM–2:00 PM	Noon
2:00–6:00 PM	Afternoon
6:00–10:00 PM	Evening
10:00 PM–2:00 AM	Night
2:00–6:00 AM	Post-midnight

Figure 1.2 shows this diurnal pattern of postings per day, which is measured as the average number of posts made per interval, over the entire 6 months history of Twitter data of the users. The mapping of 24 hours to the 6 time intervals is presented in Table 1.2. From the figure, we observe lower number of positive tweets compared to the number of negative or neutral tweets in the night and post-midnight. The number of positive tweets is higher in the morning, noon, and afternoon, but starts decreasing thereafter. In addition, we see that the overall number of tweets is highest during noon and evenings time intervals, which indicates that people are using social media more often during the day than at the end of the day. In addition, increasing number of negative tweets compared to the positive or neutral tweets at the end of the day shows that people are more likely to be negative during nights. This observation is consistent with [21], which shows worst symptoms for people who suffering from depression occur during night time.

In Fig. 1.3, the weekly pattern of postings over 7 days is presented. This pattern is measured as the average number of posts made per day over the 7 days for the entire 6 months history of Twitter data of the users. We notice that the number of negative posts is higher compared to the number of positive or neutral posts on Sundays and Mondays, since these days are the last day of weekend and the first day of weekdays which make people sad. We also see the maximum number of tweets is on Tuesdays. Furthermore, the relative number of positive tweets is highest on

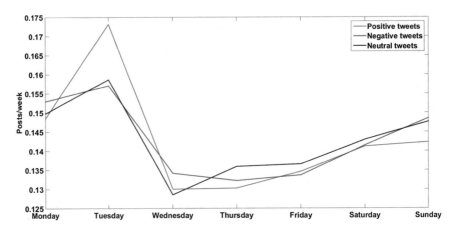

Fig. 1.3 Weekly trends of posts corresponding to all classes

Tuesdays. This indicates how much people are more positive on the second day of the week, compared to the rest of the days when the relative number of positive tweets goes lower than the number of neutral and/or negative tweets. The reason is that people are getting use to work and they are not in a weekend mood any more in the second day of week comparing Monday. We also observe that the number of posting is lowest on Wednesdays and then steadily increases until the following Tuesdays, after which there is a sudden drop.

Overall, these differences in the relative number of positive, negative, and neutral postings on different days of a week as well as over different time intervals of a day can be useful in our classifier design.

Gender Past research has shown that there has been a high correlation between gender and the way users express their emotions. In [25], women report their feeling stronger and longer, and also express them more clearly, in addition to their interest to share their emotions in public. This gender differences in expressing emotions can be considered as another feature in designing our classifier [26].

Of the 142 users in our dataset, we found 54 (about 38%) women and the rest men. Figure 1.4 compares the weekly pattern of postings for males and females. This figure clearly indicates that there are some significant differences in the posting patterns between males and females, not only in the number of postings but also in the relative numbers of positive, negative, and neutral postings. In general, we observe that females have more emotional periodic behavior during the week compared to males. This indicates how women are more used to expressing their emotions in social media compared to men who have the tendency to hide them.

Personal Activities In addition to all the features that we defined and studied so far to identify positive/negative/neutral attributes, another interesting feature is the personal concerns and activities. Personal concern is defined as a category in LIWC that contains 19 dimensions. These dimensions are related to user's occupation,

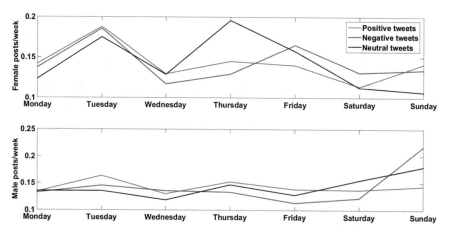

Fig. 1.4 Weekly trends of posts for both females and males corresponding to all classes

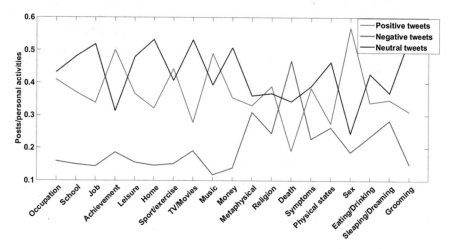

Fig. 1.5 Trends for the normalized personal concerns values corresponding to all classes

various activities like sport, sex, music, eat/drink, sleep, etc., various events such as death, and others such as metaphysical and religion. These features will be used to design our personalized classifier. We used LIWC to determine the personal concern values of each tweet. Figure 1.5 shows the normalized personal concerns values over the three categories of positive, negative, and neutral tweets.

According to this figure, the first ten activities, which are more positive ones, have lower correlation with negative tweets. For example, in "death" with a strong weight of negativity, the value of correlation with negative tweets has been increased. In contrast, positive tweets have the lower value of correlation with such an activity.

The neutral tweets follow the same pattern as either negative or positive ones, since they could not be categorized either as positive or negative. In addition to above observation, some attributes like metaphysical have almost the same value for all three categories. Considering this study, personal concerns and activities can be considered as a new and appropriate feature to identify positive, negative, or neutral attributes.

User Engagement User engagement refers to how much active a user is in Twitter. To measure user engagement, we identify two measures: number of posts made by the user in 6 months and the fraction of user's tweets that are retweets in a same time period.

User Properties We consider two user properties in Twitter: number of followers of the user and the count of her followees. Both of these properties have the potential to improve the results of our classifier.

Emoticons and Abbreviation Emoticons are extensively used by users now-a-days in their daily online activities, including Twitter. There has been some work in literature that considered the effect of emoticons in sentiment analysis. For example, in paper [15], 27 emoticons are defined as the popular, top ones used in Twitter. Some examples of emoticons contained in positive tweets are :-), :), and :p. Some examples of emoticons contained in negative tweets are :(, =(, and:-@. Instead of defining our own emoticon dictionary, we rely on LIWC, which provides the fraction of positive/negative emoticons per each tweet, and we used this value as a feature in our classifier. In addition to emoticons, since Twitter has a 140-character limit, abbreviations are extensively used. For this reason, we have used the abbreviation fraction provided by LIWC as an appropriate feature in designing our classifier.

Punctuation Another interesting feature useful in mood classification is punctuation. A punctuation category is defined in LIWC and includes 12 dimensions. Based on a similar analysis as we did for psychological dimensions, we have identified nine of these dimensions as useful in improving the performance of our classifier. These nine dimensions are: Commas, Colons, Semi-colons, Question marks, Exclamation marks, Dashes, Quotations, Apostrophes, and Parentheses.

N-Grams and Word Length n-gram including unigram, bigram, and trigram has been used. Tweets were tokenized and normalized across all of user's tweets. To improve the performance, we used some text preprocessing, including removing the common punctuation, hashtags, numbers, URLs, and non-ASCII characters. In the end, we used feature hashing to reduce the number of features as input to our learning algorithm [11]. Furthermore, due to the 140-character limit, word length and the number of words per sentence can be considered as an important feature in sentiment analysis. We have included them in our classifier design.

5 Prediction

As we mentioned before, in this paper our final goal is to implement a personalized activity-based classifier which can predict individuals' mood and emotions from their social network postings considering their temporal nature of positing over time. We first explain about our general classifier which is developed using several typical features such as n-grams and word lengths, emoticons and abbreviation, network properties, and psychological features. We will discuss how this generalized classifier results in limited performance when using activity and temporal information as extra input features. We also use these features to build a personalized classifier that predicts users' mood based on the history of their activities on articular time.

5.1 Prediction Framework

Using our labeled data, we used the supervised learning method to build a classifier to predict users' mood. To determine the best suitable classifier, we experimented with several different classification algorithms. In addition, to find the relative importance of various feature types, we experimented with separate classifiers trained on each of the features.

All features are also combined to obtain a final classifier. We compare the obtained results from different classifiers including logistic regression, naïve Bayes, and Support Vector Machine which finally found the best performing classifier using Linear SVM. In all of our analysis, we use 10-fold cross validation, with equal number of attributes as positive, negative, and neutral, over five randomized experimental runs to find the optimal results. To avoid overfitting, we also employ feature selection techniques such as Information Gain (IG) and Chi-square methods and then compare the results before and after that.

5.2 General Prediction Results

Table 1.3 shows classification results for each of the individual features as well as for all features combined before and after dimension reduction. The best results have been achieved in the reduced dimension case with 80.2% accuracy, and with high precision (80.5%) and recall (79.9%). Based on this result, we can confirm that reducing feature redundancy results in better performance compared to using all features. Among the individual features, the accuracy baseline for this dataset is 69.7% by using only unigrams. Adding bigram and trigram improved the accuracy by less than 2% as shown in Table 1.3. This observation shows that bigram and trigram cannot make a huge difference in comparison to the baseline accuracy. The main reason for this is the mostly short length of Twitter messages.

We also notice that features such as word length, emoticons and abbreviation, and punctuations do not perform well in terms of accuracy, precision, or recall.

Table 1.3 Performance metrics in mood prediction in posts using various features

1*Dataset	Precision	Recall	Accuracy
N-gram	0.739	0.715	0.715
Word length	0.41	0.415	0.401
Emoticons and abbreviations	0.419	0.452	0.45
Punctuations	0.525	0.52	0.53
Psychological features	0.667	0.648	0.648
Personal activities	0.444	0.412	0.41
User engagement	0.56	0.55	0.55
User properties	0.52	0.512	0.513
Diurnal and weekly activities	0.4	0.401	0.401
Gender	0.261	0.387	0.387
All features	0.793	0.789	0.79
Reduced dimensions	0.805	0.799	0.802

Features such as user engagement and user properties perform slightly better. However, all of these features do increase the accuracy when deployed in addition to n_gram. According to this table, psychological features provide the accuracy as 64.8% which shows a strong link between these features and the mood prediction. As we mentioned in the previous section, Fig. 1.1 shows a correlation between psychological process and positive, negative, or neutral attributes. For example, the correlation values of positive tweets and positive and affect attributes are significantly more than others. It's also clear from Fig. 1.1 that negative users tend to use negative features as anxiety, anger, and negative in their posts.

To avoid overfitting, we used IG and $\chi 2$. Table 1.4 presents the ten highest ranking features obtained and indicates how these two methods for this set of features are almost similar. This dimension reduction resulted in increasing the overall accuracy by 1.2% over all the features in the final result. Our surprise, performance of three features, namely gender, diurnal, and weekly activities is quite poor with low values of accuracy, precision, and recall. We believe that this is due to the relatively low number users in our training dataset, which is comprised of only 18 users, of which 12 are men and 6 are women. Notice that all others features are trained on Tweets whose number is quite large in the training dataset. Gender and diurnal and weekly activities increased the overall accuracy by about 1.7%.

We mentioned in the previous section about the correlation between personal activities and the three categories of mood. We noted that the negative users tend to use negative features such as death much more than positive users. To identify the best features within personal activities, and avoid overfitting, we used feature rankings. This is reported in Table 1.5. Considering these new set of features, we could increase the accuracy as much as 0.67% rather than using all personal activity's features. At the end, Table 1.3 presents the general classifier results before and after dimension reduction which is resulted from applying feature ranking on

Table 1.4 Results of psychological process feature rankings

Features	Information gain (IG)	χ2
Positive	1	1
Negative	2	3
Anger	3	2
Affect	4	4
Causation	5	5
Cognition	6	6
Optimism	7	7
Sadness	8	8
Anxiety	9	9
Feeling	10	10

Table 1.5 Results of personal activities feature rankings

Features	Information gain (IG)	χ2
Occupation	1	1
Symptoms	2	2
Leisure	3	3
Job	4	4
Physical states	5	5
Sports	6	6
Death	7	7
Sex	8	8

psychological features and personal activities. Considering these new set of features, we could increase the accuracy as much as 1.02% rather than using all features.

According to the all obtained results, we note that both the personal activities and the diurnal and weekly activities features are highly individualized for each user, and so they will be more useful in building personalized classifier rather than a generalized classifier. This is discussed in detail in the next section.

5.3   Personalized Prediction Results

Figure 1.6 shows how much personal activities can be considered as individualized features. This figure shows the correlation between personal activities and the positive, negative, and neutral tweets of three users. For example, user 1 who can be considered as positive user has high correlation with *TV/Movies* as a positive activity. This is different from users 2 and 3 who have more number of negative tweets. According to this figure, *Death* and *Sleeping/Dreaming* are activities with the highest correlation value with negative tweets. In addition, this figure indicates how much the personal activities are different for different users when they are positive or negative or neutral. For example, user 3 shows a big interest on *Sport/exercise* when she is more negative, while users 2 and 3 do not follow the

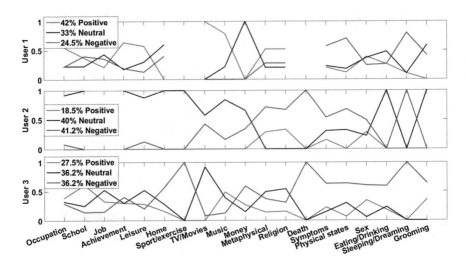

Fig. 1.6 Trends for the normalized personal concerns values corresponding to all classes for three individual users

same pattern. In addition, *Occupation* is considered as a positive activity for users 1 and 3, but not for user 2. These observations confirm the effect of personal activities to improve mood prediction via a personalized classifier. Figures 1.7 and 1.8 show the diurnal and weekly pattern of tweeting by three individual users. In Fig. 1.7, we see that user 3 has the highest number of tweets at noon, while users 1 and 2 have their highest number of tweets in evening time. Figure 1.8 shows more differences between the three users. For example, user 3 has the most number of negative tweets on Friday, compared to user 2 who has the highest number of positive tweets on this day. These observations confirm that the diurnal and weekly pattern feature could also be considered as a personalized feature, since each user uses social media in a different time interval, based on character, interest, free time, and their mood.

Based on these observations, we explored personal activities and the diurnal and weekly activities to build a personalized classifier. We selected each user as the test set and the rest of the users as the training set. We applied our general classifier for each selected user and then added the individualized features to make it a personalized classifier. Classification results are shown in Table 1.6. Here GC represents the generalized classifier discussed in the last section, T-PC refers to the personalized classifier with diurnal and weekly activities as personalized feature, A-PC refers to the personalized classifier with personal activities as personalized feature, and TA-PC refers to the personalized classifier with both diurnal and weekly activities and personal activities as personalized features.

The average accuracy for GC is 76.3%. The reason for lower accuracy compared to the one reported in the previous section is that the number of training sets here is decreased and different for each individual user. Adding the individual user-based features as diurnal and weekly pattern and personal activities increases

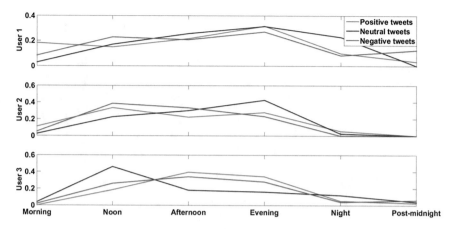

Fig. 1.7 Diurnal trends of posts corresponding to all classes for three users

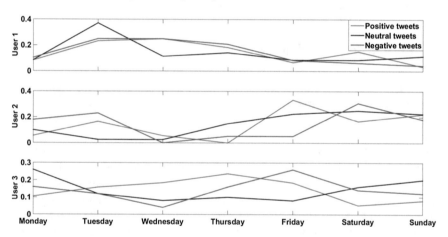

Fig. 1.8 Weekly trends of posts corresponding to all classes for three users

Table 1.6 Results of psychological process feature rankings		Accuracy
	GC	0.7632
	A-PC	0.806
	T-PC	0.797
	AT-PC	0.793

the baseline accuracy by 3.41% and 3.06%, respectively. Finally, applying TA-PC classifier resulted in 78.4%. Therefore, the best results have been achieved by activity-based classifier, with the average accuracy of 80.06%. Figure 1.9 shows the value of accuracy per user. This figure indicates how much for the most users A-PC provides the best accuracy. Applying *t*-test shows the obtained results are statistically significant by p < 0.0001.

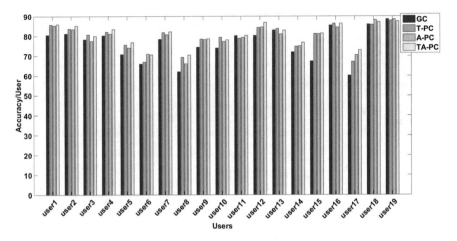

Fig. 1.9 Accuracy/user for four different classifiers

6 Conclusion and Future Work

Effect of emotion and mood on user's behavior has been studied for a long time. We have demonstrated the effect of user's activities and behavioral on mood vice versa using social media which is resulted in implementing a personalized activity-based classifier. First we employed crowdworkers recruited through Amazon's Mechanical Turk to label each tweet as positive, negative, or neutral. Then, based on the results from several experiments, we developed our general classifier using linguistic features, psychological features, social network behavioral attributes, gender, temporal nature of user postings and personal activities that provides mood prediction with high accuracy of 80.02%. Moreover, we also observed a high correlation between gender and temporal nature of user postings. We saw how women are more used to expressing their emotions in social media compared to men. In addition, women have more emotional periodic behavior during the week. Finally, we used these features in addition to personalized pattern of activities of users to build a personalized classifier that improves classification accuracy by 3.41%.

An important point to note is that negative, positive, or neutral tweets describe the sentiment of user's postings but not necessarily the mood of the user. This is because there are various cases where influential users are used as seeds to impact the activity of their followers irrespective of their actual mood as can be seen in viral marketing and terrorist networks. We believe that such cases are relatively rare, and indeed in those cases, our classifiers will merely predict the sentiments associated with the postings. Among the future directions, we hope to collect and label a higher number of specified data from highly active Twitter users with equal number of men and women including the location of their tweets. This dataset could be used to improve our personalized classifier accuracy considering some unique and differentiating attributes of users such as gender, location, and hourly nature of user postings. We are also interested in developing a recommender system to make users feel better according to their history of usage and its correlation with their emotional states.

References

1. Gouveia R, Karapanos E (2013) Footprint tracker: supporting diary studies with lifelogging. In: Proceeding of the SIGCHI conference on human factors in computing systems, CHI '13, Paris, pp 2921–2930
2. Roshanaei M, Mishra S (2014) An analysis of positivity and negativity attributes of users in Twitter. In: Proceeding of ASONAM, pp 365–370
3. Rui L, Wang S, Deng H, Wang R, Chang KC (2012) Towards social user profiling: unified and discriminative influence model for inferring home locations. In: LinkKDD. Beijing, pp 1023–1031
4. Picard R (1995) Affective Computing, M.I.T Media Laboratory Perceptual Computing Section Technical Report, vol. 321, pp 1–26
5. Damasio AR (1994) Descartes' error: emotion, reason and the human brain. Picador, Avon Books, A Division of The Hearst Corporation, New York
6. Marreiros G, Santos R, Ramos C, Neves J (2010) Context aware emotional model for group decision making. IEEE Trans Intell Syst 25(2):31–39
7. Saari T, Kallinen K, Salminen M, Ravaja N, Yanev K (2008) A mobile system and application for facilitating emotional awareness in knowledge work teams. In: Hawaii international conference on system sciences, Waikoloa, pp 1–10
8. Wilson T, Wiebe J, Hoffmann P (2008) Recognizing contextual polarity in phrase-level sentiment analysis. Comput Linguist 35(3):399–433
9. Vazire S, Gosling SD (2004) e-Perceptions: personality impressions based on personal websites. J Pers Soc Psychol 87(1):123–132
10. Back M, Stopfer J, Vazire S, Gaddis S, Schmukle S, Egloff B, Gosling S (2010) Facebook profiles reflect actual personality, not self-idealization. Psychol Sci 21(3):372–374
11. da Silva NFF et al (2014) Tweet sentiment analysis with classifier ensembles. Decis Support Syst 66(17):170–179
12. Mohammad S, Kiritchenko S, Zhu X (2013) Nrc-Canada: building the state-of-the-art in sentiment analysis of tweets. In: Proceedings of the seventh international workshop on semantic evaluation exercises (SemEval-2013), Atlanta, Georgia, USA
13. Saif H, He Y, Alani H (2012) Semantic sentiment analysis of twitter. In: Proceedings of the 11th international conference on the semantic web—volume part I, ISWC'12, Springer-Verlag, Berlin, Heidelberg, pp 508–524
14. De Choudhury M, Counts S, Gamon M (2012) Not all moods re created equal! a exploring human emotional states in social media. In: Proceeding of international AAAI conference on weblogs and social media, Dublin, pp 66–73
15. Tchokni S, Eaghdha DOS, Quercia D (2014) Emotions and phrases: status symbols in social media. In: Proceedings of the Eighth International AAAI Conference on Weblogs and Social Media Ann Arbor, pp 485–494
16. Park J, Barash V, Analytics M, Fink C, Cha M (2013) Emoticon style: interpreting differences in emoticons across cultures. In: Proceeding of international AAAI conference on weblogs and social media. Boston, pp 466–475
17. Park K, Lee S, Kim E, Park M, Park J, Cha M (2013) Mood and weather: feeling the heat?. In: Proceeding of international AAAI conference on weblogs and social media (Workshop), Boston

18. M. De Choudhury, S. Counts, E. Horvitz (2013) Predicting postpartum changes in emotion and behavior via social media. In: Proceeding of the SIGCHI conference on human factors in computing systems, New York, pp 3267–3276
19. Tausczik YR, Pennebaker JW (2009) The psychological meaning of words: LIWC and computerized text analysis methods. J Lang Soc Psychol 29(1):24–54
20. Kotikalapudi R, Chellappan S, Montgomery F, Wunsch D, Lutzen K (2012) Associating depressive symptoms in college students with internet usage using real Internet data. IEEE Tech Soc Mag 31(4):73–80
21. De Choudhury M, Gamon M, Counts S, Horvitz E (2013) Predicting depression via social media. In: Proceeding of international AAAI conference on weblogs and social media. Boston, pp 128–137
22. Park M, D. W. McDonald, Cha M (2013) Perception differences between the depressed and non-depressed users in Twitter. In: Proceeding of international AAAI conference on weblogs and social media, Boston, pp 476–485
23. Gross J (1998) The emerging field of emotion regulation: an integrative review. Rev Gen Psychol 2:271–299
24. Pennebaker JW, Mehl MR, Niederhoffer KG (2003) Psychological aspects of natural language use: our words, ourselves. Annu Rev Psychol 54:547–577
25. Fischer A, Manstead A (2000) The relation between gender and emotion in different cultures. In: Fischer A (ed) Gender and emotion: social psycholgical perspectives. Cambridge University Press, Cambridge, pp 71–94
26. Rime B, Mesquita B, Philippot P, Boca S (1991) Beyond the emotional event: six studies on the social sharing of emotion. Cognit Emotion 5:435–465

Chapter 2
Automatic Medical Image Multilingual Indexation Through a Medical Social Network

Mouhamed Gaith Ayadi, Riadh Bouslimi, Jalel Akaichi, and Hana Hedhli

1 Introduction

Online social networking is attracting more and more people in today's Internet, where users can share and consume all kinds of multimedia contents [1]. Like most people, healthcare professionals use mainstream social media networks to connect with friends and family. But almost one-third of them also join social networks focused exclusively on healthcare, where healthcare professionals can collaborate and share resources online, and patients can access more than information [2]. According to Doganay [2], patient-focused networks, often built around a particular condition or disease, give individuals and their families' supportive communities where they receive comfort, insights, and potential leads on new treatments. The data mining practices of sites like Facebook and Twitter make some patients and providers leery of posting questions or comments; and while many healthcare organizations use Facebook, Twitter, LinkedIn, Instagram, and other social tools to communicate with constituents, individuals often worry about posting information in the wrong place. By sharing data on specialized sites healthcare professionals and other users can feel safer about expressing their thoughts [2]. Franklin and Greene [3] consider that participation in the health care management can render patients longer health conscious. According to Grenier [4], the main objective behind medical networks is to foster collaboration between medical actors and to place the patient at the heart of the health system. In reality, the fact of making important decisions, related to medical images, individually, can lead the

M.G. Ayadi (✉) • R. Bouslimi • J. Akaichi
Department of Computer Sciences, ISG, BESTMOD, Tunis, Tunisia
e-mail: mouhamed.gaith.ayadi@gmail.com; bouslimi.riadh@gmail.com; jalel.akeichi@isg.rnu.tn

H. Hedhli
Emergency Department, Charles Nicolle Hospital, Tunis El Manar University, Tunis, Tunisia
e-mail: hedhli_hana@yahoo.fr

© Springer International Publishing AG 2017
J. Kawash et al. (eds.), *Prediction and Inference from Social Networks and Social Media*, Lecture Notes in Social Networks, DOI 10.1007/978-3-319-51049-1_2

physician to make errors leading to malpractices and consequently to unexpected damages. This fact is justified by a study done by The Institute of Medicine of the National Sciences Academy[1] (IMNAS) in the USA. This institute published a study estimating that up to 98,000 hospital deaths each year can be attributed to medical malpractice [5]. In order to minimize medical errors, a medical social network, as a first contribution, destined to present patients' medical images and physicians' interpretations expressing their medical reviews and advices present the solution to support collaboration between physicians and patients. This will, obviously, help to save time and better serve the patients about their situations.

Medical images and comments present the major fields among others, meant for interaction between patients and physicians. So, an attempt to index medical images content shared through the social network site becomes an important task due to the huge number of comments expressing specialist's analysis and reviews. For this reason, an analysis approach which extracts keywords and terms from comments must be used to give an overview and a summary of what exist on comments. Furthermore, extracted terms will be used to annotate and to index images in order to facilitate the search task later, through the social network site. But, we need to take into account that existing comments can be expressed in different languages because we cannot force the social network users to use one language to present their problems (for patients) or to present analysis and recommendations (for physicians). The mechanisms, used in the indexing phase, need to be adapted to the characteristics of different languages. To overcome this problem, we will present a medical images multilingual indexation based on statistical methods and on external semantic resources. This multilingual indexation's challenge has widely taken into consideration to improve the search of images later. To improve our indexing process, we added a correction of spelling mistakes using a medical vocabulary.

The remainder of this chapter is structured as follows. Section 2 presents the background needed to follow our methodology. Section 3 describes the design of our social network. Section 4 details the proposed methodology. Section 5 shows experimental measurements, discusses the experiments, and analyzes the achieved results. Finally, Sect. 6 draws some conclusions and highlights future research directions.

2   Related Work

This section on related work begins with different medical social networks main requirements and data structure, presenting the collaboration between physicians and patients. We then study work related to the main approaches dealing with an overview of images multilingual indexation firstly and some works dealing with images indexation through social networks.

---

[1] http://iom.nationalacademies.org/

2.1 Medical Social Networks

Today, social networks have the ability to connect people with just about everything. The influence of social network and those using social networks grows and changes daily, generating a profound impact on society. Furthermore, a growing majority of modern patients, particularly those with chronic conditions are seeking out social network and other online sources to acquire health information, connect with others affected by similar conditions, and play a more active role in their healthcare decisions [6]. In 2011, more than 80% of adults reported using the Internet as a resource for health care quality information and more than half of patients (57%) said they were more likely to select hospitals based on their social media presence [7]. Indeed, research shows that 81% of consumers believe that if hospitals have a strong media presence, they are likely to be more innovative than other hospitals. According to the Centers for Disease Control and Prevention (CDC), "Using social network tools has become an effective way to expand reach, foster engagement and increase access to credible, science-based health messages." [7] Medical networks have several forms which can be networks of hospital management (internal coordination and fragmentation within the hospital by specialty), resource networks (shared resources such as scanners), information networks (data collection to adjust policy information), and among many other care networks [4], which offers the members suffering from diseases an opportunity to change their lives, connect with others, and share problems. Indeed, research shows that 81% of patients believe that if hospitals have a strong media presence, they are likely to be more innovative than other hospitals.[2]

According to Feldman [7], hospitals are increasingly adopting the use of social network for a variety of key tasks, including: education and wellness programs, crisis communication, staff and volunteer training, employee and volunteer recruitment, information sharing, clinical trial recruitment and other research, public relations and marketing, etc. Since the beginning of 2011 alone, the growth in social network use for hospitals has been staggering. Ed Bennett, the manager of web operations at the University Medical Center in Baltimore, has been tracking hospital social media on his private site since 2008. He reports that as of October 2011, nearly 4000 social media sites were owned by US hospitals [7]. Many examples of medical social networks were presented in relation with different medical activities: SoberCircle[3] is intended for alcoholics and drug addicts in need for support and encouragement by others. SparkPeople,[4] Fitocracy,[5] and Dacadoo[6] do share

---

[2]http://corp.yougov.com/healthcare/

[3]http://www.sobercircle.com

[4]http://www.sparkpeople.com/

[5]https://www.fitocracy.com/

[6]https://www.dacadoo.com/

workouts and exercises in order to sustain them during weight loss. Asklepios,[7] exclusively for Canadian doctors, is meant to exchange the best practices ever and to learn from each other. CardioSource,[8] Cardiothoracic Surgery Network, concerns cardiothoracic surgery. Diabspace[9] is intended for diabetics. "Parlons Cancer"[10] is dedicated to cancer patients and their families. Renaloo[11] deals with kidney disease, dialysis, and transplantation. RxSpace[12] is dedicated to pharmacy students, pharmacists, pharmacy owners, and academia to interact with each other.

Besides, we can summarize that the top 25 health and medical social network sites are organized in three: Social networks for Doctors which offer great opportunities to confer find support and provide their own expertise, such as Sermo (http://www.sermo.com/), Ozmosis (https://ozmosis.org/) , Doctor Network (http://doctor-network.com/), etc. Social networks for nurses which can help them to connect with others who understand what is happening in the field ask and answer questions and learn more about their profession, such as Nursing Link (http://nursinglink.monster.com/), Ultimate Nurse (http://www.ultimatenurse.com/), and Nurse Zone (http://nursezone.com/). And finally, social networks for all health and medical careers, such as MedicalMingle (http://www.medicalmingle.com/), Radiolopolis (http://radiolopolis.com/), Docadoc (http://docadoc.com/), and Carenity (http://www.carenity.com/). These different kinds of social networking sites offer to their members an opportunity to be connected with others and share experiences and knowledge.

Our first goal is to design a social network dedicated to physicians and patients. The basic model of the targeted social media should take into account the management of a:

- Set of patients which provide personal information in their health care profile.
- Set of physicians providing information enabling their identifications.
- Set of mechanisms permitting to patients to upload the medical images related to their diseases.
- Set of mechanisms permitting to physicians to comment the uploaded medical images.
- Set of search functions by which patients and physicians can locate easy and efficient information about medical images.
- Site operator who controls the site and triggers a set of mechanisms permitting to collect medical images in order to process them for various purposes such as medical images' indexation.

---

[7]http://www.asklepios.com/

[8]http://www.acc.org/

[9]http://www.diabspace.com/

[10]http://www.parlonscancer.ca/

[11]http://www.renaloo.com/

[12]http://www.rxspace.com/

Like in [7], our social network, addressed to physicians and patients, can connect millions of voices to:

- Increase the timely dissemination and potential impact of health and safety information;
- Leverage audience networks to facilitate information sharing;
- Personalize and reinforce health messages that can be more easily tailored or targeted to particular audiences;
- Empower people to make safer and healthier decisions;
- Facilitate interactive communication, connection, and public engagement;
- Updates patients about changes in physician's practice;
- UKeeps patients informed about upcoming appointments, tests, immunizations;
- Engages patients in discussions about key health issues;
- Answers patients' medical questions;
- Communicates with family members, other caregivers;
- Grows physician's practice;
- etc.

In our work, we respect that our social network needs to contain all of these features, situated above.

2.2 Multilingual Indexation Approaches

2.2.1 An Overview

Annotation, indexing, and retrieval of image content in the large scale online repositories have become an increasingly active field. Annotation and tagging have been recognized as a very important and essential mechanism to enable the effective organization and sharing of large scale of images information. Therefore, efficient automatic annotation and tagging methods are highly desirable. This interdisciplinary research direction has attracted various attentions and resulted in many algorithmic and methodological developments. In the literature, different works of extracting terms from textual corpus use two approaches: statistical analysis and the linguistic or structural analysis. Statistical analysis is based on the study of the contexts of use and distributions of terms in documents. Linguistic analysis uses of language knowledge, such as morphological and syntactic structure of terms. Other works combine the two approaches and constitute an approach called "hybrid or mixed approach."

Several studies [8–11] have used Semantic Resources (RS) in the process of indexing. They find that improving information retrieval systems is based on such resources for indexing multilingual corpus or for detecting lemmatization. They use UMLS (Unified Medical Language System) to index ImageCLEFmed collection. Similarly, thesaurus MeSH (Medical Subject Heading) is used in [11] to index documents of TREC. The same steps are usually used; Starting by language identification, subsequently the extraction of terms by a linguistic method adapted to the language, finally the projection of extracted terms on semantic resource to detect concepts.

Maisonnasse et al. [8] use a linguistic method of concepts' detection, performed in the collection of ImageCLEFmed 2007. This method was used with three different linguistic tools: MetaMap[13] is a morphosyntactic analyzer associated with UMLS which can extract concepts from documents. This analyzer deals only with written in English. MiniPar[14] can extract terms in English also. TreeTagger[15] is an analyzer which detects the grammatical category of a word and its lemma. There is a version of TreeTagger for English, French, and German. Although a large number of image tags can be generated, also, in short time, these approaches depend on the availability of human annotators and are far from being automatic. Similarly, research in the other direction via text-to-image synthesis [12–14] has also helped to harvest images, mostly for concrete words, by refining image search engines.

The indexing method used in these works consists of the same steps: identification of the language of the document, extracting terms of this language by a linguistic method adapted to the language, detection of concepts by projecting the extracted terms on semantics resource. Unlike these methods, we would like to show that with an external semantic resource of sufficient quality, it is not necessary to use linguistic tools suited to a given language and a method of extracting purely statistical terms allows obtaining results of equivalent quality. Thus, with this statistical method, we will not have to change language analyzer whenever the document language changes.

2.2.2 Indexation Approaches via Social Networks

The idea behind an "ideal" annotation is to provide recommendations of annotations based on collaborations between users. Several online systems have sprung into existence to achieve annotation of real-world images through human collaborative efforts (Flickr) and stimulating competition [15, 16], in the context of ESP game project. From the same point of view, Google sets up Google Labeler Image a game which consists in determining the common directions of the images for two distant players. This kind of work neglects that the common directions between the users is, in our opinion, strongly related to a social aspect and a preliminary knowledge. The number of images on Facebook, as an example, has exceeded 60 billion by the end of 2010, and around 138 MB of new content is being uploaded every minute. This user-uploaded and user-generated audio-visual content belongs to the established concept of user-generated content (UGC) [17]. UGC includes all kinds of data that come from regular people who voluntarily contribute with data, information, or

---

[13]https://metamap.nlm.nih.gov/

[14]http://ai.stanford.edu/~rion/parsing/minipar_viz.html

[15]http://www.cis.uni-muenchen.de/~schmid/tools/TreeTagger/

media that then appears before others in a useful or entertaining way. Automatic tagging and search for image content has been a tremendous challenge, particularly in uncontrolled environments such as UGC applications. Collaborative annotation and tagging has been a typical and promising approach for tagging of user-generated multimedia content [17]. A limited number of studies using data from a social network to improve the suggested annotations are highly recommended. For the purpose of this study, it is essential to integrate or benefit from the use of the following works such as Shevade et al. [18], Stone et al. [19], Messaoudi et al. [5], and Bouslimi et al. [20].

Most approaches to automatic image annotation have focused on the generation of image labels using annotation models trained with image features and human annotated keywords [21–24]. Instead of predicting specific words, these methods generally target the generation of semantic classes (e.g., vegetation, animal, building, places, etc.), which they can achieve with a reasonable amount of success. Recent work has also considered the generation of labels for real-world images. In the same context, Shevade et al. [18] combine measurements of similarity between users. These measurements understand proximities between users in the social network, semantic similarities between concepts using ConceptNet, and of the similarities between events. The annotations are generated while using (1) annotations of the most similar user in the social network, (2) the similarity of the images based on their contents, and (3) the application of the activation spreading on the graph the most similar concepts. The activation spreading is a process of research initiated by the labeling of a set of nodes source with weights. The spreading with the other nodes is iterative, and it takes into account the relation between the nodes.

Zunjarwad et al. [25] have taken the work of Shevade et al. [18]. The contribution of Zunjarward et al. is summarized in what they call social network of confidence. This approach is based on two measures. The first one is founded on a binary value attributed manually by the user to each member of his social network. This binary value corresponds to a value of confidence. However, the second measurement is based on the co-occurrence of the user with others in the same event. The events are tags attributed by the users. A value of confidence is granted to a user, if he has annotated photos with an event that matches the current event annotation.

Stone et al. [19] proposed a solution to identify people on Facebook using the recognition and similarity of faces by content. To suggest the figurants' user names, they estimate statistically the intensity of inter-user relationship. To proceed this method, one should consider two metrics which are: the number of photos of a user on which a person is identified or the pictures of his friends and that the number of pictures when user is present together with that person.

Manning et al. [26] have proposed a collaborative approach in which patients seek via a social network to find quick analysis of their medical images by expert doctors. In addition, they use a mixed approach terminology extraction annotations. Once the terms found, they seek in MeSH thesaurus the concepts that are related to the keywords already found.

Fuming et al. [27] have also proposed a collaborative approach, wherein several different statistical models are effectively combined to predict the annotation for each image. In addition, they combined the correspondence between keywords tokens and visual image/regions and word-for-word correlation to improve the annotation.

Sun et al. [28] have combined the annotation of similar images via collaborative approach. Similar images are searched with search engines, and their tags then infer via word correlation the annotation of target image.

Bouslimi et al. [20] have suggested an automatic system for medical image annotation that combines textual and visual descriptors using latent semantic vector to build a semantically medical image. Indeed, to automatically annotate a new medical image, they compare the vector describing the source image with the vectors that are already existing in the database.

Messaoudi et al. [5] have proposed a collaborative approach, in which patients seek via a social network to find quick analysis of their medical images by expert doctors. In addition, they use a mixed approach terminology extraction annotations. Once the terms found, they seek in MeSH thesaurus, the concepts that are related to the keywords already found.

Kanishcheva and Angelova [29] have suggested an approach to image auto-tagging refinement. They have presented a post-processing tool that refines tags associated with images. The tool works stepwise in two phases. There are several steps in phase 1 that maintain tags as English wordforms. Step 1.1 cleans mistakes (as tags sometimes contain typos) and splits words which are written together. Step 1.2 finds plural forms and transforms them to singular; it also analyzes the inflection. Step 1.3 analyzes synonyms. Step 1.4 analyzes phrases. Step 2, about semantic analysis that consolidates tags using English language resources like WordNet, will be considered below.

Bouslimi and Akaichi [30] have proposed a social network for collaboration whose goal is learning among medical residents that are currently in the course of their training in radiology. Indeed, the terminology extraction from comments has allowed them to obtain the relevant terms and using a correlation with the MeSH thesaurus in order to have access to get the concepts. They used the keyword results for the construction of an annotation of the medical image.

We have made a comprehensive review on the state of the art of medical images' indexation. The automatic extraction tools help terminologists to validate the extracted terms. We note the existence of a wide variety of extraction tools. Linguistic techniques have a fine linguistic description and have the ability to handle small corpus. However, it requires a large linguistic knowledge. The statistical approaches have the advantage of requiring no prior knowledge of language and are applicable on corpus for which no external resource (dictionary, stop list, ontology, etc.) has been developed. The results of statistical approaches are strongly related to the corpus studied and cannot be generalized beyond this context. These approaches are efficient for corpus sufficiently large size. They are not applicable on small corpus sizes. Finally, the hybrid approaches, which provide quality results, present a compromise between statistical methods and linguistic methods. The idea of combining these two methods is relevant. Indeed, this combination takes advantage of the fine linguistic analysis and robustness digital analysis. The hybrid approach takes advantage of the speed and independence from the field of statistical methods. Indeed, Harrathi [31] demonstrated that the use of methods of extracting purely statistical terms provides results of equivalent quality. For author, it is not necessary to use linguistic tools adapted to a given language. He shows that with an external semantic resource of sufficient quality, statistical approach gives greater results than those using linguistic techniques. In addition, the statistical approach is simple to implement as opposed to linguistic approaches. Our approach is similar to that of Manning et al. [26], and inspired from the work of Bouslimi and Akaichi [30] and Messaoudi et al. [5], through which we propose a social network for the collaboration between physicians and patients. But, most of these works do not cover all languages. They present a monolingual terminology extraction based on the extraction of terms in a specific language: the French language or the English language. We will present, in our work, a medical images multilingual indexation from comments, based on statistical methods and on external semantic resources. Our approach allowed us to obtain the relevant multilingual terms. After that, we get the multilingual concepts by using the correlation with the multilingual MeSH thesaurus. In addition, the multilingual keywords results construct an annotation of the medical image, in different languages. In what follows, we explicitly describe the description of our social network, firstly and our indexation's approach, secondly. We added a correction of spelling mistakes using the medical vocabulary to improve our indexing process, inspired from the work of Bouslimi and Akaichi [30].

3   Social Network Architecture Description and Implementation

Social network is the study of social entities, as a group of people linked to one another by one or more common attributes, and their interactions and relationships. The interactions and relationships can be represented with a network or graph, where each node represents an actor and each link represents a relationship. A social network plays an important role as a support for the dissemination of information, ideas, and impact among its members. It may be a major selling point for patients looking for physicians. In fact, social networking has been proven effective in sharing knowledge and establishing communication among patients and physicians.

Our proposed social network, as many others described in the literature Gong and Sun [32] and Almansoori et al. [33], aims to allow patients and physicians to connect with each other by eliminating all geographical and time frontiers. Both users exploit it to seek advices and to share experiences related to medical images' interpretations and diseases' analysis.

Like in Facebook, patients and physicians need to be registered by creating profiles about themselves (detailed information). This step is very important in order to use our medical social network. Our social network site is a virtual place where registered patients upload their medical images to be commented by various registered physicians. Not only registered patients can upload their medical images, but even registered physicians have this right, especially medical students, in the objective to learn from this collaboration. Uploading and sharing images are not the only functionalities. Users can also sharing status, medical events, and any information related to medical activities. The main objective, behind using our medical social network, is to enable patients and physicians to exchange information, share knowledge, and experiences. It, also, gives to its members the opportunity to connect and communicate with others in need of support and encouragement related to different situations and health care problems. Consequently, the patient's condition represents the heart of the health system justified by experts' interpretations and analysis expressed by comments. Patients frequently place trust in peer recommendations on social network site making them key platforms to influence change. We used PHP (Hypertext Preprocessor), in the development of our social network, and MySQL for the database management. PHP is a server-side scripting language not only designed for web development but also used as a general-purpose programming language. It ensures many features that allow creating and managing an entire social network website. Pages were written in HTML, PHP, Ajax, and JavaScript and they were designed using Dreamweaver. In order to better explain our medical social functionalities, we describe in the following some interfaces of the social network:

- The first interface to use is the identification interface. Patients and physicians need to provide the registration process, as new users. The registration interface dedicated to physicians has more specificity because profiles have different structures.
- After the identification, the basic functions and services of the social network are displayed in the main interface. Patients and physicians can perform different activities, especially, get access to posting, commenting functions, and uploading images, etc.
- Mixed posts and comments, in different languages, performed by different patients and physicians are entered and displayed, in the posting interface. Posts will be treated by back office functions implementing our multilingual indexation's approach.
- To answer urgent questions and/or advices, synchronous communications between members (patients and physicians, physicians and physicians) are performed in the communication interface. Mixed posts performed by different physicians and patients, according to their opinions and questions, are displayed in the posting interface.

The different interfaces of our social network allow an easier mapping for users, their intent, and targeted functions. The most important is that these interfaces describe the way of interaction between patients and physicians with the social network site and the way of interaction between patients and physicians with each other by exploiting existing functions. The social network content blocks are visually separated. This separation will help to organize pages content and each element is well defined and presented separately, to succeed the interaction between patients and physicians. In fact, this separation makes the content understandable, easy to recognize and makes the content more reachable. In order to improve the interaction between users (patients and physicians, physicians and physicians), our social network contains an advanced search function, allowing the organization of the connections between patients and physicians having common interests related to diseases and their analysis, interpretations and treatments. This function supports users to rapidly find the content and contacts they are searching for. Considering the recommendations of a design expert, the created interfaces are simple in terms of color scheme and graphics, for example, in Facebook. The idea consists of using a few colors and the background is generally white. This management of color scheme helps the exploitation of our social network site by physicians and patients, makes the content well presented, and comments better seen and readable. Therefore, buttons and links are placed almost on every page of the social network. Some links are related to the navigation processes and some others permit to users to regulate specific functions. Buttons are used to associate users to actions and to navigate among different pages; they are clear and more remarkable.

When incoming comments or messages appear on the social network sites, physicians and patients need to react to them in real time. For this reason, they should consider establishing their own social network presence. So, we assure our social network with a synchronous communication, in order to establishing interaction and sharing information. We, also, equipped our social network with a real-time update feature ensuring the delivery of updates (medical images uploading, physicians' annotations, etc.) as soon as they are submitted. In order to encourage patients and physicians to exploit our social network site, many actions could be performed such as suggesting new friends, preferably in relation with medical activities, interests, events, and groups. This is performed to extend their social circles allowing implicitly the extension of the shared knowledge about medical images and the associated diseases.

The extraction of more knowledge, from various posts or comments, is possible by using mining tools. For this reason, our social network is equipped by mining tools such as those presented in Xie et al. [34], where authors consider the emergence and pervasively of online social networks that have enhanced web data with developing interactions and communities both at large scale and in real-time.

The successful building of our social network needs to respect security aspects and to protect users' privacy. Physicians must protect their own privacy, as well as that of their patients. To do it, we use privacy settings on our social network site to keep out everyone except patients or fellow physicians. We will also consider security aspects such as those presented in [35], where the author proposed methods to secure healthcare social networking sites providing users with tools and services to easily establish contact with each other around shared problems and utilize the wisdom of masses to outbreak disease.

The successful building of the social network site needs also to respect a privacy policy considered to protect the users whether they are patients or physicians. The American Medical Association [36] recently adopted a policy to ensure professionalism and protect patients' privacy during social network activities. According to the AMA policy (AMA 2012), physicians should:

- Never post identifiable patient information;
- Separate personal from professional content;
- Use privacy settings when going on social networking sites, to protect personal information;
- Monitor their sites and their presence on other Internet sites, to make sure content is appropriate and accurate.

The AMA policy also warns that anything doctors post on the Internet may become public and permanent. The AMA [36] explains inappropriate postings can have a detrimental effect on a doctor's reputation among patients and colleagues. Because such postings can also be harmful to the medical profession, the AMA advises physicians also to be alert for unprofessional content placed on the Internet by other physicians. And if doctors see something, they are obligated to do something about it (AMA 2012). In the same time, physicians must protect their own privacy, as well as that of their patients. Once doctors have a presence on the Internet, though, privacy may be difficult to maintain. They must exercise caution about the information they put on social network sites—both their own and others. Moreover, it is essential that physicians avoid sites that could prove awkward if their visit there were revealed [36].

4   The Proposed Methodology

Figure 2.1 shows a structure of the conceptual design for our multilingual approach which we consider as a mixed method involving the combination of statistical methods with a multilingual external semantic resource. From Fig. 2.1, we can see that our methodology mechanism has four main components: pre-processing unit; cleaning, correcting, and lemmatization unit; terms' extraction unit, and concepts' detection unit.

Each of these steps, as shown in Fig. 2.1, consists of several separate processes: cleaning, lemmatization, multilingual simple terms' extraction, multilingual compound terms' extraction, and multilingual concepts' extraction. Our method of multilingual concept's detection is a preliminary step to a semantic indexing process. Our terminology extraction approach is a mixed approach which combines the two approaches linguistically and statistically. The algorithm, describing the steps of the proposed method in generally, is below.

As described by this algorithm, the multilingual indexation's approach initiates by the creation of the textual corpus containing comments extracted from the medical social network and performed on a medical image. Then, we will describe the different steps of our approach step by step.

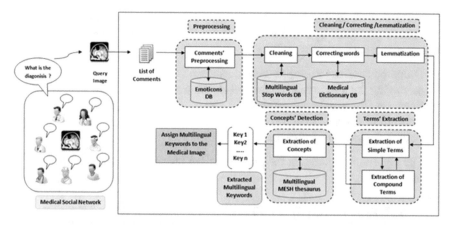

Fig. 2.1 Proposed methodology pipeline

MULTILIGUAL INDEXATION'S APPROACH ALGORITHM

INPUT

C_i : A Comment in the Corpus of comments *{Comment}*

OUTPUT

MC_{on} : Multilingual Concepts *{List of full Multilingual Keywords}*

VAR

MS_t : Multilingual Simple Terms
MC_t : Multilingual Compound Terms
M_{esh} : Multilingual MeSH Thesaurus
L_w : A List of Multilingual Words

0. **BEGIN**
1. **PREPROCESS OF** C_i {Comment} **IN** List of Comments
2. $L_w \leftarrow$ **CLEAN** C_i {Comment} **USING** a Multilingual Anti-Dictionary;
3.   **FOR EACH** W_i {word} **IN** L_w **DO**
4.   **CORRECTING** (W_i) **USING** a Multilingual Medical Dictionary;
5.   **PERFORM** Lemmatization (W_i);
6.   **END FOR**;
7. $MS_t \leftarrow$ **EXTRACT** Multilingual Simple Terms (L_w);
8. $MC_t \leftarrow$ **EXTRACT** Multilingual Compound Terms (MS_t, L_w);
9. $MC_{on} \leftarrow$ **EXTRACT** Multilingual Concepts (MS_t, MC_t, M_{esh});
10. **END.**

4.1 Comments' Pre-processing

The first step consists of the pre-processing operation that removes punctuation such as (,;?;!,[],()Etc..) and cuts the sentence into words. We remove also the emoticons' symbols, by using an anti-dictionary that contains the emotions that are used frequently in twitter, Facebook, etc. Our pre-treatment algorithm contains five procedures, one for each sub-step. The algorithm will be presented below contains a procedure to replace the emoticon's symbol and the punctuation space by character space, then a procedure that convert text to lowercase, procedure that Clean Text and finally procedure Split text into words. So, the pre-processing step is composed of five preliminary stages:

- Decomposition of the corpus sentences.
- Removing the punctuation points and cleaning sentence.
- Removing the emoticons' symbols.
- Converting sentence to lowercase.
- Cutting the sentence into words.

The algorithm, describing the pre-processing phase of the proposed method, is below.

PRE-PROCESSING ALGORITHM

INPUT
   C_i : A Comment in the Corpus of comments *{Comment}*

OUTPUT
   L_{word} : List of Words

VAR
   S_{en}: A Sentence in the list of sentences
   Low_s: Lowercased Sentence

0. **BEGIN**
1. **FOR EACH** C_i {Comment} **IN** List of Commentary **DO**
2.   **DECOMPOSITION OF** C_i {Comment} **INTO** Sentences;
3.   **FOR EACH** S_{en} {Sentence} **IN** Sentences **DO**
4.   **DELETE** the emoticons' symbols **SINCE** emoticons' DB;
5.   **REMOVING** the punctuation points;
6.   **CLEANING** (S_{en});
7. Low_s ← **LOWERCASE** (S_{en});
8. L_{word} ← **CUTTING** (Low_s) **INTO** words;
9.   **END FOR;**
10. **END FOR;**
11. **END.**

4.2   Cleaning, Correcting, and Lemmatization

4.2.1   Cleaning

The cleaning step is about removing the stop words. A stop word (empty word) is a word which is not useful to index a document such as prepositions, articles, pronouns, some adverbs and adjectives, and finally some frequent word. Each language possesses a list of stop words which corresponds to his specificity. This type of word has a low informative power. To assure this task, we use a multilingual anti-dictionary containing the blackest words, in different languages, that seem useless in the medical field. We cite among them, from three different languages:

(many, how, again, which, since then, this, on some, but there, like why, however, when, which, soon, etc.) in English; (alors, au, aucuns, aussi, autre, avant, avec, avoir, bon, car, ce, etc.) in French, and (aber, als, am, an, auch, auf, aus, bei, bin, bis, bist, da) in German. The rest of the words (full words) will be used to describe the images contents because it has a high discriminatory power.

4.2.2 Correcting Words

Comments may include many misspelled words. These misspelled words may have a negative effect on our indexation's approach and on the search function later. For this reason, correcting words is based on using of a medical dictionary to detect misspelled words. A word is considered misspelled when it does not appear in the dictionary. In this case, we presented to use the LEXILOGOS[16] Dictionary of medical terms. LEXILOGOS can cover many languages especially from the European countries. There are many mathematical metrics providing a measure of similarity between two character sequences. According to Bouslimi and Akaichi [30], the use of the two distances Levenshtein [37] and Jaccard [38], to correct spelling mistakes and to find the nearest word to be corrected, is highly efficient. The performance of the two distances has been proven, also, by Heasoo et al. [39]. According to Navarro [40], Levenshtein distance can be used to generate possible corrections. This distance is equal to the number of characters one has to delete, insert, or replace to go from one sequence to another. In most cases, the distance between a misspelled word and its correction is 1. All candidate corrections will be made up of all words obtained by:

- Inserting a character in the misspelled word;
- Deleting a character of the misspelled word;
- Substituting a character of the misspelled word.

The formula expressing the Levenshtein distance is the following:

$$lev_{a,b}(i,j) = \begin{cases} \max(i,j) & \text{if } \min(i,j) = 0, \\ \min \begin{cases} lev_{a,b}(i-1,j)+1 \\ lev_{a,b}(i,j-1)+1 \\ lev_{a,b}(i-1,j-1)+1_{a_i \neq b_j} \end{cases} & \text{otherwise} \end{cases} \qquad (2.1)$$

We will present, thereafter, two examples of operations they need to correct English and French words.

Example 1 (Lev hemoragiie, Hémorragie)

h	é		m	o	r	r	a	g	i			e
	replace				insert			replace	delete			
h	e		m	o	r		a	g	i		i	e

---

[16]http://www.lexilogos.com/medical_dictionnaire.htm

Example 2 (Lev vontrycular, Ventricular)

v	e	n	t	r	i	c	u	l	a	r
	replace				replace					
v	o	n	t	r	y	c	u	l	a	r

Like in [30] and for more precision, we used the Jaccard distance to support the Levenshtein distance. The Jaccard distance is the ratio between the cardinality (size) of the intersection of the sets considered and the cardinality of the union of the sets. It evaluates the similarity between the sets. The words w1 and w2 are represented, not as vectors, but as letter sets. The similarity obtained is $d_{jaccard}(w1, w2) \in [0, 1]$. The formula expressing the Jaccard distance is the following:

$$d_{jaccard}(w_1, w_2) = \frac{\vec{w_1}.\vec{w_2}}{|\vec{w_1}||\vec{w_2}| - \vec{w_1}.\vec{w_2}} \tag{2.2}$$

The choice of the most probable correction is done by attributing to each correction candidate a score. The higher the score is, the more likely it is. That is to say, the candidate correction is the correct spelling of the word to correct [30].

4.2.3   Lemmatization Words

In other side, comments may include different forms of words belonging to various families, for grammatical reasons. Lemmatization, that seeks the canonical form of words, is the process of grouping together the different inflected forms of a word so they can be analyzed as a single item: the name, the plural, the verb in the infinitive, etc. It is used to regroup words in their belonging family. To find the lemmas, we implement the stemmer algorithm which seeks the root (prefix) and then assigns the suffix for a parent noun. According to Frakes and Fox [41], stemming algorithms are used in many applications related to natural language processing such as text analysis systems, information retrieval, and database search systems. We update and use Porter algorithm [26] which is considered as the best known manner corresponding to our needs. Table 2.1 presents an example of rules for English language from different steps of the Porter stemmer.

Table 2.1 An example of rules

Example of rules	Example of results
Rule1: sses → ss	Caresses → caress
Rule2: ator → ate	Operator → operate
Rule3: ness → to remove	Goodness → good
Rule4: ible → to remove	Defensible → defens

We implemented the entire algorithm using all grammatical rules associated with French, English, German, Spanish, and Italian languages, in order to validate our method. The algorithm, describing the Cleaning, Correcting, and Lemmatization phase of the proposed method, is below.

4.3   Terms' Extraction

We distinguish two types of terms: simple terms composed of a single word and the compound terms composed of a sequence of words. We used a multilingual anti-dictionary, containing multilingual stop words, for extracting simple terms and a statistical measure for the extraction of compound terms.

CLEANING CORRECTING AND LEMMATIZATION ALGORITHM

INPUT
    W_i : A World in the List of Words

OUTPUT
    L_{lfw} : A List of Lemmatized Corrected Words

VAR
    ST_w: A Stop World in the Stop Worlds DB
    W_{md}: A Word in the Medical Dictionary
    d_1: Levenshtein Distance Value
    d_2: Jaccard Distance Value

0. **BEGIN**
1. **FOR EACH** W_i {World} **IN** List of Words **DO**
2.   **FOR EACH** ST_w {Stop World} **IN** Stop Worlds DB **DO**
3.   **IF** W_i **IS A** ST_w **THEN**
4.   **PERFORM CLEANING** (W_i);
5.   **ELSE IF** W_i **IS NOT A** ST_w **THEN**
6.   **FOR EACH** W_{md} **IN** Medical Dictionary **DO**
7. $d_1 \leftarrow$ **Calculat_Levenshtein**(W_i, W_{md});
8. $d_2 \leftarrow$ **Calculat_Jaccard** (W_i, W_{md});
9.   **CORRECTING** $(W_i, W_{md}, \mathbf{MAX}(d_1, d_2))$;
10. $L_{lfw} \leftarrow$ **LEMMATIZE** (W_i);
11.   **END IF**;
12.   **END FOR**;
13. **END FOR**;
14. **END**.

4.3.1   Simple Terms' Extraction

According to Bouslimi and Akaichi [30], many approaches seek to define a key term based on some statistical features and studying their relation with the notion of importance of a candidate term. If a candidate is considered the important term in a document analyzed, then it is relevant as a key term. TF-IDF of Jones [42] and Likey of Paukkeri and Honkela [43] are two methods which compare the behavior of a candidate term in the analyzed document with his behavior in a collection of documents (reference corpus). The objective is to find candidate terms whose behavior in the document that varies positively compared to their global behavior in the collection. The two methods were expressed by the fact that a term has a strong importance face to face of the analyzed document if there is present, then it is not in the rest of the collection. According to the same study, TF-IDF gives good results compared to Likey, and that is why in the extraction step of the keywords we have used this method. This method combines two factors, the local weighting (TF) which quantifies the local representation of a term in the corpus of comments and the overall weighting (IDF) which measures the global representative of the term on the collection of corpus based on provided comments. We will keep only the terms that the output value exceeds, according to a threshold fixed to 0.125. The formula expressing the measure is the following:

$$TF - IDF(term) = TF(term) * \log \frac{N}{DF(term)} \qquad (2.3)$$

where TF represents the number of occurrences of a term in the analyzed document, DF represents the number of documents in which it is present, and N is the total number of documents. The higher the score of TF-IDF of a candidate term, the more it is important in the analyzed document. The algorithm, describing the simple terms' extraction step of the proposed method, is as follows:

SIMPLE TERMS' EXTRACTION ALGORITHM

INPUT
 W_i : A World in the List of Lemmatized Words
 C_i : A Comment in the Corpus of comments
OUTPUT
 L_{st} : A List of Simple Terms

CONSTANT
 $T_{hrd} = 0.125$ // The Threshold
VAR
 W_{eit} : float // The Weighting of The Word in The Comment

0. **BEGIN**
1. **FOR EACH** C_i {Comment} **IN** List of Commentary **DO**
2. **FOR EACH** W_i {World} **IN** List of Lemmatized Words **DO**
  // Calculate The Weighting of The Word
3. $W_{eit} \leftarrow TF(term) * \log\left(\frac{N}{DF(term)}\right)$
4.   **IF** $W_{eit} > T_{hrd}$ **THEN**
5. $L_{st} \leftarrow W_i$;
6.   **END IF**;
7. **END FOR**;
8. **END FOR**;
9. **END**.

4.3.2   Compound Terms' Extraction

This step is about the identification if the term is a compound or a single word. A collocation is a combination of recurrent words which are more often found together than each one of them alone. Some collocations are fixed noun phrases that are specific to a domain [31]. Mutual Information, as a measure, is used to extract relevant collocations. This measure compares the probability of co-occurrence of words and the probability of each words separately [5]. The process of extracting complex terms is iterative and incremental. We build complex terms of length $n + 1$ words from the words of length n. For each sequence of words, we compute the value of the *MI*. If the sequence of words exceeds the threshold set to 0.15 in our case, the sequence will be comprised on the list of compound terms. Note that the computation of *MI* is ensured by the following formula:

$$MI(m_1, m_2) = \log_2 \frac{P(m_1, m_2)}{P(m_1) * P(m_2)} \tag{2.4}$$

where $P(m_1$ and $P(m_2)$ are an estimation of the probability of occurrence of the words m_1, m_2 and $P(m_1, m_2)$ is an estimation of the probability that the two words appear together. The algorithm, describing the compound terms' extraction step of the proposed method, is as follows:

COMPOUND TERMS' EXTRACTION ALGORITHM

INPUT
  L_{st} : A List of Simple Terms
  L_{word} : List of Words
OUTPUT
  L_{ct} : A List of Compound Terms

CONSTANT
  $T_{hrd} = 0.15$ // The Threshold
VAR
  MI: The Mutual Information
  W_i : A World in The Corpus
  t_i: A Term in The List of Simple Terms
0. **BEGIN**
1.   **FOR EACH** t_i {term} **IN** L_{st} **DO**
2.   **FOR EACH** W_i {World in The Corpus} **IN** L_{word} **DO**
3.   MI $\leftarrow log_2 \frac{P(t_i, W_i)}{P(t_i) * P(W_i)}$
4.   **IF** MI > T_{hrd} **THEN**
5. $L_{ct} \leftarrow$ (CONCAT(t_i, W_i));
6.   **END IF**;
7.   **END FOR**;
8. **END FOR**;
9. **END.**

4.3.3 Concepts' Extraction

After extracting simple and compound terms which participate to candidate keywords, it comes to the extraction of concepts which are chosen from a controlled
vocabulary (a dictionary, thesaurus, or a list of terms, etc.). This is a verification
step which comes to use an external semantic resource. We use, in our case, a
multilingual Mesh thesaurus to cover different languages. We extract concepts by
projecting those terms on the thesaurus. More precisely a medical multilingual
thesaurus MeSH is used to filter keywords obtained in the previous step, in order
to verify that the extracted term belongs, or not, to the medical vocabulary. We used
the UMLS (Unified Medical Language System) provided by the National Library
of Medicine. It is a multilingual meta-thesaurus covering the medical field. This
resource was created to facilitate research and the integration of information from
multiple electronic sources of biomedical information [44]. It is the fusion of several
semantic resources. UMLS contains more than one million of concepts related to
more than 5.5 million terms in 17 languages. The 17 languages are not covered
in the same way in UMLS. English is the language most represented with 68% of
the vocabulary, German vocabulary covers 2.84%, and French vocabulary covers
2.55%. The algorithm, describing the concepts' extraction step of the proposed
method, is as follows:

CONCEPTS' EXTRACTION ALGORITHM

INUPUT
    L_{st} : A List of Simple Terms
    L_{ct} : A List of Compound Terms
OUTPUT
    L_{con} : A List of Concepts

VAR
    S_{ti} : A Term in The List of Simple Terms
    C_{ti} : A Term in The List of Compound Terms

0. **BEGIN**
1.   **FOR EACH** S_{ti} {simple term} **IN** L_{st} **DO**
2.   **IF** S_{ti} **IN** {Thesaurus MeSH} **THEN**
3. $L_{con} \leftarrow S_{ti}$;
4.   **END IF**;
5.   **END FOR**;
6.   **FOR EACH** C_{ti} {compound term} **IN** L_{ct} **DO**
7.   **IF** S_{ti} **IN** {Thesaurus MeSH} **THEN**
8. $L_{con} \leftarrow C_{ti}$;
9.   **END IF**;
10.   **END FOR**;
11. **END.**

5   Experimental Results

After developing the medical social network site, the challenge is to extract relevant multilingual terms and keywords from comments to index and annotate images.

5.1   Data Test and Evaluation Criteria

Experiments have been carried out to validate the efficiency of the proposed model. The experiments were carried out on a Core i_3, 2.4 GHz processor with 4 GB RAM using NetBeans editor. The Multilingual Anti-Dictionary is saved in an XML file. The files will be examined using JDOM (Java). For querying the MESH[17] thesaurus (meshdata.rdf) file, we used the java API Sesame of Thomas Francart which allows executing applications written in SPARQL queries.

The relevance of our approach was evaluated on preselected medical images, which are supplied by internal physicians and residents during their training, diagnostic, and images analysis in relation with patients' states and clinical cases, at Charles Nicolle Hospital in Tunis. This collection contains 200 medical images annotated by 40 physicians, involved in various specialties, from different countries. Different analyses are the results of the collaboration between Tunisian physicians and other physicians from countries like France, UK, and Germany. We collected 100 examination cases in different languages, where each examination consists on a comment related to a medical image. Analyses are essentially written in French, English, Italian, Spanish, and German.

We also perform experiments on the collection of ImageCLEF'2013. This collection is composed of over 45,000 biomedical research articles in PubMed Central (R). Each document is constituted of a medical image and a portion of text. The collection contained over 300,000 images including MR CT, PET, ultrasound, and combined modalities in one image. The images are very heterogeneous in size and content.

We evaluate the performance of this approach by using three measures frequently used in the evaluation of information retrieval systems, namely the precision, the recall, and the MAP (Mean Average Precision) metrics. The MAP represents the quality of a system based on different levels of precision [26]. The precision is the number of correct concepts divided by the total number of extracted concepts. The equation of the precision is

$$Precision = \frac{The\ number\ of\ correct\ concepts}{The\ total\ number\ of\ extracted\ concepts} \qquad (2.5)$$

---

[17]http://www.nzdl.org/Kea/download.html

The recall is the ratio between the correct terms and the total number of correct concepts that should have been extracted. The equation of the recall is

$$Recall = \frac{The\ number\ of\ correct\ concepts}{The\ total\ number\ of\ correct\ concepts\ that\ should\ have\ been\ extracted}$$
(2.6)

5.2 Evaluation and Results of Our Approach

Our study has been started with a pre-processing phase of the annotations where we have used our own algorithm that cleans the stop words and emoticons' symbols in the text. We used the indexation's method to lemmatize comments and to extract relevant medical terms intended to be used for medical images annotation and indexation. A medical dictionary is used that contains all vocabularies used in medicine to correct errors found during the indexation. This correction is based on the use of the combination of the two distances of Levenshtein and Jaccard to evaluate the correlation between the correct term and the terms found in the medical dictionary. We attempt then to prove that results are confirmed in practice. Figure 2.2. presents an example of comments on cranial CT scan, performed by six physicians. Indeed, according to this example, we notice the existence of an ambiguity of using different languages to express diagnostic and analysis and another ambiguity of the existing of incorrect words.

Table 2.2 shows the calculation that has been done during the correction phase, in relation with multilingual terms existing in Fig. 2.2.

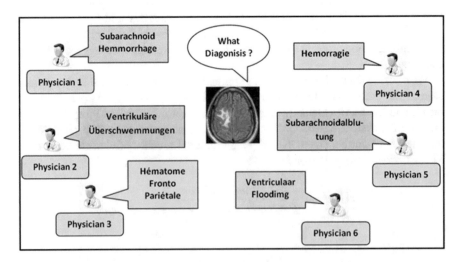

Fig. 2.2 Example of cranial CT scan commented by physicians

Table 2.2 Table of correct words

List of words	Levenshtein distance	Jaccard distance	Max (d_J, d_L)	Correct words
Hemorragie	0.85	0.65	0.85	Hémorragie
Ventrikuläre Überschwemmungen	1	1	1	Ventrikuläre Überschwemmungen
Hématome FrontoPariétale	1	1	1	Hématome Fronto Pariétale
Ventriculaar floodimg	0.92	0.88	0.92	Ventricular flooding
Subarachnoidalblutung	1	1	1	Subarachnoidalblutung
Subarachnoid hemmorrhage	1	1	1	Subarachnoid hemmorrhage

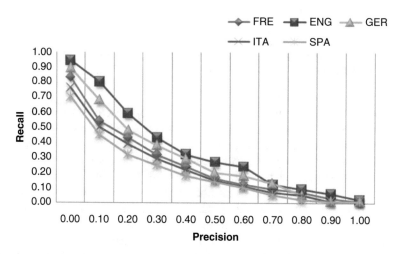

Fig. 2.3 Precision versus recall curves for English, German, French, Italian, and Spanish languages

We perform some experiments to check the effectiveness of the proposed method. The application of the precision-recall curve enables the evaluation of the proposed scheme. It is clear, after many tests, from Fig. 2.3. that proposed method works very well with five different languages: French, English, Italian, Spanish, and German. Figure 2.3 shows the curves of average precision to 11 points of recall with our method of concepts' detection. The curves show that the coverage of the language has a direct impact on system's performance. Indeed, we find that the accuracy obtained for English language is larger than other languages. UMLS covers English better than other languages. The average precision obtained for French and German languages are almost similar with a slight improvement for German language. These two languages are covered in the same manner in UMLS. German language has a slightly higher coverage than French language in UMLS. The average precision obtained for Italian and Spanish languages are almost similar. These two languages are also covered in the same manner in UMLS.

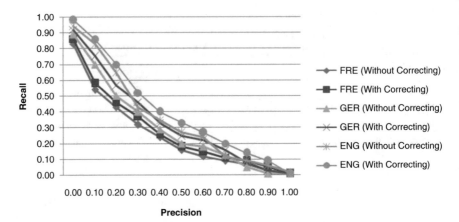

Fig. 2.4 Precision versus recall curves for English, German, and French languages with and without the spelling correction phase

In order to evaluate the performance of our approach, Fig. 2.4. presents a comparative study of our approach firstly without taking into consideration the spelling correction phase and secondly with the spelling correction phase. The spelling correction phase has a positive effect on the indexation process.

We report experimental results that show the feasibility and utility of the proposed algorithm and compare its performance with state-of-the-art methods: we conducted a comparative study with three linguistics methods presented by Maisonnasse et al. [8]. The comparison is carried out by an application of each method and record different results from tests, performed on our dataset. We conducted, also, a comparative study with the works of Messaoudi et al. [5] and Bouslimi and Akaichi [30]. Authors, in these works, handle the same dataset. It allows us to compare our results with published results. Figure 2.5 shows that our proposed system performance is better than three linguistics methods presented by Maisonnasse et al. [8]. We take into account the comparison of our approach with and without the spelling correction phase.

Figure 2.6. shows that our proposed system performance is also better than the other systems presented by Messaoudi et al. [5] and Bouslimi and Akaichi [30].

Table 2.3 presents the performance of each system using the hybrid approach in its indexing and we record results from 299,764 words annotated manually. Figure 2.7 shows the comparison of the proposed system with other systems in terms of the Precision and the Recall. It shows the Precision and the Recall values of each system via different values from Table 2.3 by a vertical bar.

Table 2.4 shows that our proposed system performance is better than all systems in term of Mean Average Precision, after many tests. Figure 2.8 shows the comparison of the proposed system with other systems in term of Mean Average Precision. It shows the Mean Average Precision value of each system via the value from Table 2.4 by a vertical bar.

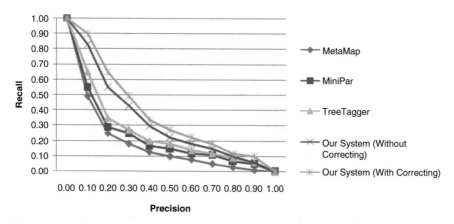

Fig. 2.5 Precision versus recall curves for the comparison between our system (with and without the spelling correction phase) and MetaMap, MiniPar, and TreeTagger systems

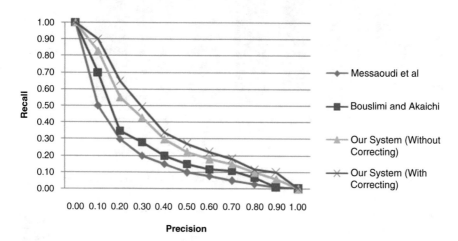

Fig. 2.6 Precision versus recall curves for the comparison between our system (with and without the spelling correction phase) and Messaoudi et al. and Bouslimi and Akaichi systems

For more precision, we evaluated the performance of our solution with that of Messaoudi et al. [5] and Bouslimi and Akaichi [30] by using 2 corpus of CRTT,[18] which are composed by articles taken from the base Science Direct. We see from the figures below, our solution provides good results by the subscribers to the results obtained by Messaoudi et al. [5] and Bouslimi and Akaichi [30].

---

[18]http://perso.univ-lyon2.fr/~maniezf/Corpus/Corpus_medical_FR_CRTT.htm

Table 2.3 Precision and recall of different systems

System	Automatically extract	Corrects	Precision	Recall
MetaMap	263,465	168,631	0.64	0.56
MiniPar	258,752	170,893	0.66	0.57
TreeTagger	264,533	185,269	0.70	0.61
Messaoudi et al. [5]	265,674	191,556	0.72	0.63
Bouslimi and Akaichi [30]	261,181	190,662	0.73	0.63
Our system (without correction)	265,820	207,340	0.78	0.69
Our system (with correction)	265,820	215,315	0.81	0.71

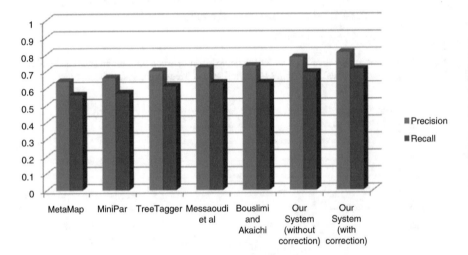

Fig. 2.7 Precision/recall comparisons between the proposed system and existing systems

Table 2.4 Average precision comparison between the proposed system and existing systems

	MAP
MetaMap	0.246
MiniPar	0.246
TreeTagger	0.258
Messaoudi et al. [5]	0.262
Bouslimi and Akaichi [30]	0.269
Our system (without correction)	0.272
Our system (with correction)	0.276

The extracted multilingual keywords and concepts are directly attributed to the medical image. The use of the multilingual thesaurus MeSH to extract medical concepts constitutes a positive point. The selected multilingual terms shall be defined such as keywords to automatically annotate the medical image through the medical social network site (Figs. 2.9 and 2.10).

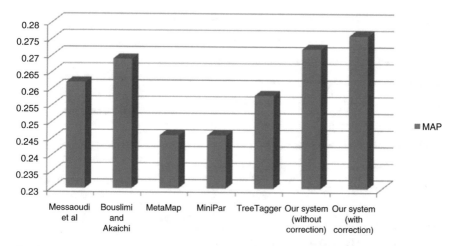

Fig. 2.8 Mean average precision comparisons between the proposed system and existing systems (with and without the spelling correction phase)

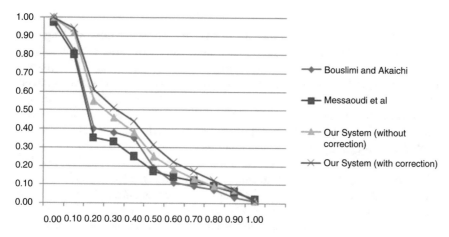

Fig. 2.9 Precision versus recall curves for corpus "transfCli-bio-txt"

6   Conclusion and Future Work

Social image indexing and retrieval in the large databases of the social networks advanced the challenges to form a new problem that needs special handling. Comments present a source for the indexation of images existing in the social network site. Firstly, we proposed in this paper a medical social network destined to both medical images presented by patients and physicians' analyses expressing their medical recommendations and advice. Secondly, this study has presented a new approach to automatic multilingual annotation of medical images based on comments from the medical social network site. This approach has the primary

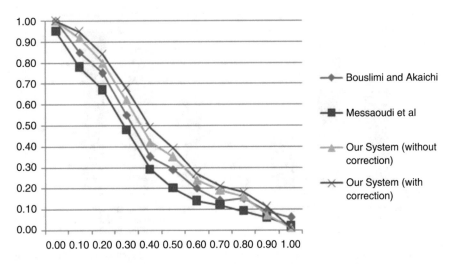

Fig. 2.10 Precision versus recall curves for corpus "AnnCardAng-txt"

goal of removing the ambiguity that comments are expressed in different languages. Our approach is about the extraction of multilingual terms and current concepts existing on comments, in order to facility the search task later. This approach focuses on algorithm mainly based on statistical methods and an external semantic resource. Statistical methods used to select important and significant multilingual terms from comments. Then an extraction of concepts step through a wealthy external multilingual medical semantic resource by a multilingual medical thesaurus (UMLS). Through the extraction of concepts, completeness is verified because it seeks to ensure that those selected keywords are the most complete possible. We worked also, in this paper, to remove the ambiguity of misspelled words in comments. The combination of the two metrics of Levenshtein and Jaccard is mainly the basic technique used in this study. We used a multilingual medical dictionary that contains the different vocabularies used in medicine. We evaluated with experiments on real annotations on medical images, which are obtained from the Hospital Charle-Nicolle in Tunis with collaboration between physicians from other countries. The results have shown the relevance of our approach, especially with European language. There are some limits in relation with other existing languages that need a specific treatment like Chinese and Arabic languages. Future work will focus on presenting a specific treatment for other existing languages in order to enlarge the circle of users of our social network site.

References

1. Zhi W, Wenwu Z, Peng C, Lifeng S, Shiqiang Y (2013) Social media recommendation. In: Social media retrieval. Computer communications and networks. Springer, Berlin. doi:10.1007/978-1-4471-4555-4 3
2. Doganay S (2014) Healthcare social networks: new choices for doctors, Patients. Available from http://www.information.com/healthcare/patient-tools/healthcare-social-networks-new-choices-for-doctors-patients/d/d-id/1234884
3. Franklin V, Greene S (2007) Sweet talk: a text messaging support system. J Diabetes Nurs 11(1):22–26
4. Grenier C (2003) The role of intermediate subject to understand the structuring of an organizational network of actors and technology case of a care network. In: Proceedings of the 9th conference of the association information and management, Grenoble
5. Messaoudi A, Bouslimi R, Akaichi J (2013) Indexing medical images based on collaborative experts reports. Int J Comput Appl (0975-887) 70(5):1–9
6. Daniel RG, Liza SR, Jennifer LK (2013) Dangers and opportunities for social media in medicine. Clin Obstet Gynecol 56(3). doi:10.1097/GRF.0b013e318297dc38
7. Feldman DL (2012) Medical social media networks: communicating across the virtual highway. Q J Health Care Practice Risk Manag Infocus 18(1):2–5
8. Maisonnasse L, Gaussier E, Chevallet J-P (2009) Combination of semantic analysis to search for medical information. In: RISE (Research Information semantics) within the INFORSID' conference, Toulouse
9. Gaussier E, Maissonnasse L, Chevallet JP (2008) Multiplying concept sources for graph modeling. In: CLEF 2007. LNCS 5152 proceedings, pp 585–592
10. Lacoste C, Chevallet JP, Lim j-h, Wei X, Raccoceanu D, Hoang D, Vuillenemot F (2006) Ipal knowledge-based medical image retrieval in imageCLEFmed 2006. In: Working notes for the CLEF 2006 workshop, Alicante
11. Neil S, Velte T, Jie H, Wei Z, Clement Y (2007) Knowledge intensive conceptual retrieval and passage extraction of biomedical literature. In: 30th annual 66 international ACM SIGIR conference on research and development in information retrieval
12. Li L-J, Fei-Fei L (2009) Optimol: automatic online picture collection via incremental model learning. Int J Comput Vis 88(2):147–168
13. Collins B, Deng J, Li K, Fei-Fei L (2008) Towards scalable dataset construction: an active learning approach. In: Proceedings of the European conference on computer vision
14. Mihalcea R, Leong C-W (2009) Towards communicating simple sentences using pictorial representations. Mach Transl 22:153–173
15. Von Ahn L, Dabbish L (2004) Labeling images with a computer game. In: Proceedings of the SIGCHI conference on human factors in computing systems, Vienna. ACM, New York, pp 319–326
16. Truran M, Goulding J, Ashman H (2005) Co-active intelligence for image retrieval. In Proc. of the 13th annual ACM international conference on multimedia, Hilton. ACM, New York, pp 547–550
17. Li Q, Lu SCY (2008) Collaborative tagging applications and approaches. IEEE Multimed 15(3):14–21
18. Shevade B, Sundaram H, Xie L (2007) Modeling personal and social network context for event annotation in images. In: Proceedings of the 7th ACM/IEEE-CS joint conference on digital libraries, Vancouver, BC. ACM, New York, pp 127–134
19. Stone Z, Zickler T, Darrell T (2008) Autotagging facebook: social network context improves photo annotation. In: Proceedings of the 1st IEEE workshop on internet vision (CVPR 2008), p 8
20. Bouslimi R, Messaoudi A, Akaichi J (2013) Using a bag of words for automatic medical image annotation with a latent semantic. Int J Artif Intell Appl 4(3):51

21. Barnard K, Forsyth D (2007) Learning the semantics of words and pictures. In: Proceedings of international conference on computer vision
22. Jeon J, Lavrenko V, Manmatha R (2007) Automatic image annotation and retrieval using cross-media relevance models. In: Proceedings of the ACM SIGIR conference on research and development in information retrieval
23. Makadia A, Pavlovic V, Kumar S (2008) A new baseline for image annotation. In: Proceedings of the European conference on computer vision
24. Wang C, Blei David, Fei-Fei Li (2009) Simultaneous image classification and annotation. In: Proceedings of the IEEE conference on computer vision and pattern recognition
25. Zunjarwad A, Sundaram H, Xie L (2007) Contextual wisdom: social relations and correlations for multimedia event annotation. In: Proceedings of the 15th international conference on multimedia, Augsburg. ACM, New York, pp 615–624
26. Manning CD, Raghavan P, Schütze H (2008) Introduction to information retrieval. Cambridge University Press, New York
27. Fuming S, Yong G, Dongxia W, Xueming W (2010). A collaborative approach for image annotation. In: PSIVT, 2010, image and video technology, Pacific-Rim symposium on, image and video technology, Pacific-Rim symposium on 2010, pp 192–196. doi:10.1109/PSIVT.2010.39
28. Sun F, Ge Y, Wang D, Wang X (2010) A collaborative approach for image annotation. In: Proceedings of the PSIVT'10. IEEE Computer Society 2010, Singapore, pp 192–196. ISBN:978-0-7695-4285-0
29. Kanishcheva O, Angelova G (2015) A pipeline approach to image auto-tagging refinement. In: BCI '15 proceedings of the 7th Balkan conference on informatics conference, New York, NY. doi:10.1145/2801081.2801108
30. Bouslimi R, Akaichi J (2015) Automatic medical image annotation on social network of physician collaboration. Netw Model Anal Health Inform Bioinforma 4:10. doi:10.1007/s13721-015-0082-5
31. Harrathi F (2010) Extraction de concepts et de relations entre concepts à partir des documents multilingues: approche statistique et ontologique. PhD Thesis, INSA Lyon
32. Gong J, Sun S (2011) Individual doctor recommendation model on medical social network. In: Proceedings of the 7th international conference on advanced data mining and applications (ADMA'11)
33. Almansoori W, Zarour O, Jarada TN, Karampales P, Rokne J, Alhajj R (2011) Applications of social network construction and analysis in the medical referral process. In: Proceedings of the 2011 IEEE ninth international conference on dependable, autonomic and secure computing (DASC'11)
34. Xie Y, Chen Z, Cheng Y, Zhang K, Agrawal A, Liao WK, Choudhary A (2013) Detecting and tracking disease outbreaks by mining social media data. In: Proceedings of the twenty-third international joint conference on artificial intelligence (IJCAI'13)
35. Li J (2014) Data protection in healthcare social networks. J IEEE Softw 31(1):46–53
36. AMA Policy (2012) Professionalism in the use of social media. American Medical Association, 2012 Annual meeting. http://www.ama-assn.org/ama/pub/meeting/professionalism-social-media.shtml
37. Levenshtein VI (1966) Binary codes capable of correcting deletions, insertions and reversals. Sov Phys Dokl 10:707–710
38. Jaccard P (1901) Distribution de la flore alpine dans le Bassin des Drouces et dans quelques regions voisines. Bull Soc Vaud Sci Nat 37(140):241–272
39. Heasoo H, Lauw Hady W, Getoor L, Ntoulas A (2012) Organizing user search histories. IEEE Trans J Mag Knowl Data Eng 24:912–925
40. Navarro G (2001) A guided tour to approximate string matching. ACM Comput Surv 33(1):31–88
41. Frakes WB, Fox CJ (2003) Strength and similarity of affix removal stemming algorithms. In: Newsletter of ACM SIGIR forum homepage archive, vol 37(1), New York, pp 26–30
42. Jones K (1972) A statistical interpretation of term specificity and its application in retrieval. J Doc 28(1):11–21

43. Paukkeri M, Honkela T (2010) Likey: unsupervised language-independent keyphrase extraction. In: Proceedings of the 5th international workshop on semantic evaluation, Uppsala, Sweden, pp 162–165
44. NLM (2009) NLM unified medical language system fact sheet. Available from: http://www.nlm.nih.gov/pubs/factsheets/umls.html.Cited23/04/2009

Chapter 3
The Significant Effect of Overlapping Community Structures in Signed Social Networks

Mohsen Shahriari, Ying Li, and Ralf Klamma

1 Introduction

Link prediction and community detection have been center of much attention for a long while [27, 31]. Social networks are changing over time and some links (dis)appear, some relationships are (de)formed and some people leave/join communities [4]. Link prediction has applications in recommender systems. In fact, link prediction approaches are employed to recommend people to users of digital media [22]. The link prediction problem tracks the evolution of graph structures and predicts (de)formation of links [15]. Besides to the evolution property, networks have dense components named communities. Communities are structures in which density inside them is higher than the density among them [31]. There exists broad research on disjoint community detection algorithms. In disjoint communities, members only belong to one community at a time. Disjoint approaches are not any longer realistic and research has concentrated on overlapping community detection (OCD) approaches. In overlapping communities, each node may belong to more than one community and actively play roles in them [34]. In other words, disjoint community detection algorithms are artificial to be applied on real world networks. For instance, a person is a student at a university and also joins a sport club simultaneously, therefore this person is overlapping among communities.

In several social networks, communications among members denote positive and negative meaning. A positive connection shows trust and friendship. On the contrary, a negative link denotes enmity or distrust. These connections can be inferred explicitly or implicitly. For instance, in Wikipedia, one can vote positively

M. Shahriari (✉) • Y. Li • R. Klamma
Advanced Community Information Systems (ACIS), RWTH Aachen University, Ahornstr. 55, 52056 Aachen, Germany
e-mail: shahriari@dbis.rwth-aachen.de; yingli@dbis.rwth-aachen.de; klamma@dbis.rwth-aachen.de

© Springer International Publishing AG 2017
J. Kawash et al. (eds.), *Prediction and Inference from Social Networks and Social Media*, Lecture Notes in Social Networks, DOI 10.1007/978-3-319-51049-1_3

or negatively toward a candidate. Moreover, positive and negative sentiments of posts in forums can be implicitly mapped to signed networks [21, 25]. Overlapping community structures also exist in networks with both positive and negative connections. Density of connections is important like unsigned graphs, however, balancing theory is significant to identify communities, too. In other words, inside communities tend to be positive and among communities tend to exist negative connections. In this regard, overlapping communities raise a couple of questions such as how dense are overlapping parts of communities? Do overlapping nodes carry broader view of the network? How can we figure out significance of overlapping community structures and overlapping members?

To answer these questions, first we need an OCD algorithm that is better to be fast, have good performance and simple logic. Hence, we propose an algorithm based on two simple social dynamics named signed disassortative degree mixing and information diffusion for signed networks. To identify significance of overlapping members, we need to think of a way to explore the overlapping community structures. However, effect of overlapping nodes have rarely been investigated and there exist few works regarding their importance in social networks [49]. In this paper, we connect the OCD problem with the sign prediction problem in order to investigate the importance of overlapping nodes.

To further motivate our work, we mention recommendation systems and open source developer networks. These systems can benefit from significance of overlapping members. For instance, in informal learning networks, overlapping members can be recommended to experts while these people can contribute to scale up communities [19]. Not only learning networks but also open source developer communities may require overlapping nodes to extend the community of developers and contribute to the longevity and success of the project. Consequently, we are inspired to identify overlapping members in signed networks. Moreover, we apply overlapping members to the case of sign prediction problem in order to observe their effect in building prediction models. As results indicate, overlapping community structures possess significant information to predict signs of links.

1.1 Contribution of the Paper

In this paper, we propose a two-phase leader-based OCD approach suitable for signed graphs named SDMID [37]. In the first phase, we identify most influential nodes in the network. To identify leaders, we compute both effective degree and disassortative value of a node. Effective degree also considers the effect of negative links and assumes a node is promoted when it receives positive links and is degraded when it takes negative ones. Not only effective degree of nodes is important to identify most influential nodes but also disassortative degree mixing is significant. In fact, among high effective degree nodes, we select those satisfying the disassortative degree mixing property. Disassortative degree mixing is a sign of dissimilarity with the neighbours of a node. Real world networks like Internet and co-authorship networks possess this property and communities appear around disassortative high impact vertices [30, 42, 43]. Details of the first phase are explained in the method section. In the second phase, we define a cascading behaviour tuned for signed networks based on network coordination game in unsigned graphs. In the defined cascade, nodes positively connected to their neighbours obey their opinions and nodes negatively connected to their neighbours disobey them. Cascades initiated by identified leaders overlap and have different sizes. Details of the cascading process are demonstrated in the method section. Results indicate that SDMID achieves better performance regarding modularity and frustration.

Furthermore, we investigate how much significant overlapping nodes are and how much information they possess in comparison to other nodes. As for this, we differentiate overlapping nodes from intra and extra nodes of communities. We employ the sign prediction problem to make models based on simple in-degree and out-degree features of trustor and trustee. In other words, each of incoming and outgoing nodes' neighbour of trustor and trustee can fit in one of intra, extra and overlapping categories. Additionally, we extend original HITS and PageRank to overlapping community-based HITS (OC-HITS) and overlapping community-based PageRank (OC-PageRank). OC-HITS and OC-PageRank differentiate among intra, overlapping and extra nodes in their updating rules through α, β and γ coefficients. Actually, ranking algorithms connect the OCD with the sign prediction problem. Results indicate that overlapping nodes play an important role and their prediction accuracies are competitive to the intra and extra categories. Results open up avenues to investigate our method on other social networks, for example, unsigned social networks.

In summary, the paper makes the following contributions:

- A local two-phase OCD algorithm named SDMID is proposed for signed social networks. We compare SDMID with two other baselines named SPM and MEAs-SN. Results indicate better performance of SDMID regarding modularity and frustration.
- We explore the significance of overlapping nodes in the sign prediction problem. To this end, we also extend classical HITS and PageRank algorithms to overlapping community-based HITS (OC-HITS) and overlapping community-based PageRank (OC-PageRank) to differentiate among intra, overlapping and extra community members. Part of community detection and sign prediction results were published at proceedings of ASONAM and i-KNOW conferences [37, 39].

The rest of the paper is organized as follows. In Sect. 2, we discuss the related work. In Sect. 3, we bring the needed terms and notations to understand the paper. In Sect. 4, we propose SDMID, extended HITS and PageRank algorithms. Afterwards, we demonstrate the features and classifiers for sign prediction in Sect. 5. In Sect. 6, we introduce real world and synthetic networks and evaluation metrics. Next, we elaborate the results in Sect. 7. Finally, we conclude and discuss the future works in Sect. 8.

2 Related Work

Community detection is part of the network analysis to obtain a better understanding of complex networks. Communities usually contain elements which are more similar and tight in comparison to the rest of the network. There has been global and local approaches to find community structures in unsigned networks. Among the global approaches of community detection, algorithm by Newman et al. is notable [32]. Due to shortcomings of global approaches, for example, resolution limit or high running times, researchers tend to investigate local methods of community detection based on clustering coefficients, random walk and leader-based approaches [7, 17, 26, 38, 42, 48]. According to the network types, research on community detection can also be categorized into unsigned and signed networks. Early research mainly focused on unsigned networks. These approaches fall under two basic categories: graph partitioning and graph clustering. Graph partitioning is dominated by the bisection method proposed by Pothen et al. [35], which is based on the eigenvectors of the graph Laplacian and iteratively divides groups into two subgroups. The weakness of such algorithms lies in their inability to divide a graph in an arbitrary number of parts [9, 31]. Kernighan and Lin proposed a heuristic which improves an initial and arbitrary clustering of nodes by optimizing the number of intra-community and inter-community edges with a greedy algorithm [18]. Girvan and Newman proposed a divisive algorithm, which focuses on the edges most possibly located between communities and detects communities by removing those edges from the network [32]. They also introduced the concept of modularity, which evaluates the quality of the obtained community structure [32]. Afterwards, a series of algorithms based upon modularity have been developed; examples could be simulated annealing developed by Guimera et al. [14] and spectral optimization proposed by Wang et al. [46].

Algorithms in unsigned networks, however, cannot be applied directly to the signed networks, because the balancing theory requires edges within communities to be positive while those between communities to be negative [10]. The balancing theory has found its way into a number of studies. To find communities in signed networks, Doreian et al. proposed a frustration criteria which considers both the friendship inside communities and hostilities among them. This error value increases when the links violate the balancing theory. They find optimal communities by a local search algorithm [11]. Yang et al. have developed an agent-based algorithm. They assume that it is more probable for a node to walk inside its community than to get across the community border [50]. By examining localized aggregated transition probabilities, communities are detected. In another work, modularity was changed to be adapted to the case of correlated data which is suitable for signed graphs and to find communities [13]. Shen used generative models to find regular structures in the networks. Similar patterns of the networks are identified via using statistical inference and expectation maximization [41]. Wu et al. propose a spectral analysis approach using cluster orthogonality in cluster eigenspace [47]. The idea behind this algorithm is mapping nodes from node to sphere coordinate and then using k-means clustering [47]. Among spectral approaches, Anchuri and Magdon-Ismail propose a two-phase spectral approach mixed with optimization equations to find the covers [3]. Anchuri and Magdon-Ismail also came up with a two-step approach to maximize modularity and minimize frustration [2]. In their algorithm, all nodes will be initially assigned to two communities according to the leading eigenvector of generalized modularity matrix. Those nodes with low absolute value in the eigenvector will then be reassigned if their movement can lead to a higher overall value covering both modularity and frustration. This division will be repeatedly carried out until the overall value cannot be improved any more. The multi-objective approach of Amelio and Pizzuti aims at simultaneously optimizing two objectives [1]. One is to raise the density of positive intra-connections and reduce the density of negative inter-connections while the other one is to minimize both negative intra-connections and positive inter-connections. A single best solution is then obtained by choosing either minimum frustration or maximum modularity [1].

However, the above algorithms designed for signed networks all generate hard partitions, i.e., a node can only belong to one single community. Till now there has been only scattered research addressing this deficiency. Chen et al. have developed the signed probabilistic mixture model (SPM) which is based on expectation-maximization method. This algorithm generates positive and negative links with different probabilities and can produce a soft partition of a network, which is based on the probability of a node belonging to each community [8]. Limits of this model include the prerequisite of prior knowledge about the total community number and its inability to deal with directed graphs. Liu et al. proposed a multi-objective evolutionary algorithm based on similarity for community detection in signed networks (MEAs-SN). The community structure is identified by maximizing the sum of positive similarities within communities as well as the sum of negative similarities between communities. This algorithm can identify both separate and overlapping communities. But like the SPM model, it can only handle undirected networks [28].

Predicting future relationships in social networks have already gained lots of attention. Sign prediction is one special form of link prediction when we limit ourself to predict -1 or $+1$. Regarding sign prediction, kernel methods and machine learning approaches are notable [21, 25]. In machine learning approaches, one would extract some features to the case of binary classification. Leskovec et al. applied signed triads and degree as features and used a logistic regression classifier to predict signs of links [25]. Shahriari and Jalili [36] applied optimism and reputation features to predict signs of links. Dubois et al. use probabilistic path probability and network distance for the train and test phase of sign prediction [12].

Regarding sign prediction based on community detection approaches, Symeonidis et al. apply a spectral clustering approach to the case of link prediction [44]. They use eigenvalues and eigenvectors of Laplacian matrix for partitioning the data and perform a link prediction problem. Their approach is suitable for unsigned networks. Javari and Jalili use clustering to the case of sign prediction [16]. They apply cluster-based collaborative filtering for the sign prediction [16]. Shahriary et al. [40] apply community detection algorithm in signed networks to the case of link prediction. To the best of our knowledge, we are the first to use overlapping community detection and overlapping nodes to the case of sign prediction problem. We would like to observe to which extent overlapping nodes are effective in comparison to intra and extra nodes.

3   Use of Terms, Variables and Definitions

In this section, we define the terms and variables used in this work. Signed social networks are graphs $G(V, E)$ in which V is the set of nodes and E is the set of edges that take two possible values of -1 and $+1$. We denote nodes which are pointing to node i, by $indeg(i)$ and nodes in which node i points to them by $outdeg(i)$. Because we have positive and negative links, we indicate $indeg^+(i)$ and $indeg^-(i)$ as nodes who positively and negatively vote toward node i. Respectively, we define $outdeg^+(i)$ and $outdeg^-(i)$ as set of nodes which node i positively and negatively points to them. The set of neighbours of node i is represented with $Nei(i)$. Alternatively, we represent $Nei(i) = deg(i) = indeg^-(i) \cup indeg^+(i) \cup outdeg^-(i) \cup outdeg^+(i)$ as the set of nodes adjacent to i.

The sign prediction is an extension of the link prediction problem in which we build a probabilistic model to reliably predict sign for hidden edge from a trustor to a trustee. We denote trustor by u and trustee by v. Community detection algorithms look for more connected components named communities. These algorithms usually perform based on finding more dense parts of the graph. Real world networks comprise nodes which are members of different communities and can actively play role in them. We denote the covers found by a community detection algorithm by $C = \{C_1, C_2, \ldots, C_k\}$ in which k is the number of communities and these communities overlap. In other words, for communities C_i and C_j, $C_i \cap C_j \neq \emptyset$. Additionally, we extend the in-degree notation for node i to $indeg_{type}^{sign}(i)$ which sign can have values of $+$ and $-$. $type$ can get three values of $intra$, $extra$ and Ovl. For example, for a sample node j, $indeg_{Ovl}^+(j)$ indicates the set of in-degree nodes not only positively link to node j but are also overlapping among identified communities. Similarly, $indeg_{intra}^-(j)$ is the set of nodes that negatively point to node j and are as well in the same community in which node j belongs to. Respectively, we can extend it to the extra and outgoing case by mentioning that $outdeg_{extra}^-(j)$ is the set of nodes in which node j negatively points to them and are member of different communities than j belongs to (they are not in the same cover). To denote the size of sets, we use #, i.e., $\#outdeg_{Ovl}^+(j)$ is the number nodes in which node j positively vote to them and these nodes are also overlapping among communities.

Node ranking is a well-used problem in social networks that assign rank values to nodes based on their connections with the neighbours. Usually ranking algorithms consider an iterative process of updating a rank vector. Among classical ranking algorithms, HITS [20] and PageRank [33] are noticeable. HITS considers two vectors of hub (h_u) and authority (a_u). These vectors are initialized with some random values and updated until convergence ($a_i = \sum\limits_{j \in indeg(i)} (h_j)$ and $h_i = \sum\limits_{j \in outdeg(i)} (a_j)$). Moreover, PageRank employs a random walker on the adjacency graph of connections and assigns each node a stationary value. PageRank initializes the rank vector with some random values and updates the vector until convergence $PR_i = \zeta \times \sum\limits_{j \in outdeg(i)} (\frac{PR_j}{\#Outdeg(j)}) + (1 - \zeta) * (\frac{1}{\#V})$. ζ is a teleporting parameter in order to avoid stopping the walk.

4   Signed Disassortative Degree Mixing and Information Diffusion Approach

In this section, we demonstrate the proposed two-phase OCD approach.

4.1   Identifying Leaders

In the first phase, most influential nodes named leaders are identified. Leaders not only possess high influence but also they have high dissimilarity values with their neighbours. Due to the existence of negative links, computing influence is different from unsigned networks. We define the influence or effective degree of node i as follows:

$$ED(i) = \frac{\#indeg^+(i) - \#indeg^-(i)}{\#indeg^+(i) + \#indeg^-(i)}, \qquad (3.1)$$

where $ED(i)$ is the effective degree of node i. We consider nodes with positive ED values to identify leaders. Moreover, a leader not only has high degree influence but also it has degree difference with its neighbours. Disassortative degree mixing denotes heterogeneity and disassortative-ness of a node indicates how much the node differs from its neighbours' degree. One can define the normalized disassortative-ness of node i as follows:

$$DASS(i) = \frac{\sum\limits_{j \in Nei(i)} (\#deg(j) - \#deg(i))}{\sum\limits_{j \in Nei(i)} (\#deg(j) + \#deg(i))}, \qquad (3.2)$$

where $DASS(i)$ is the disassortative-ness of node i. To compute this value, we consider negative edges as positive and only take into account the number of connections. Now, with contribution of $DASS$ and ED, one can further define the local leadership of each node i as follows:

$$LLD(i) = \alpha \times DASS(i) + (1 - \alpha) \times ED(i), \tag{3.3}$$

where LLD is the local leadership degree of node i, α is a parameter which weight to either disassortative-ness or influence of node i. In the experiments, we set the α to 0.5 because both effective degree and disassortative degree mixing are two significant factors to identify leaders and it is better not to underestimate one of them. Moreover, we proceed to identify local leaders. To this end, we compare local leadership value of each node to its neighbours. In other words, a node i is a local leader if for $\forall j \in Nei(i)$

$$LLD(i) > LLD(j), \tag{3.4}$$

We denote the local leader set with LL. Among the local leaders, those owning sufficient number of followers are global leaders. Hence, a node i is a global leader if

$$LL(i) > AFD, \tag{3.5}$$

in which AFD is the average follower degree and counts for the average number of followers. AFD can be computed as follows:

$$AFD = \frac{\sum_{i \in LL} |FL(i)|}{|LL|}, \tag{3.6}$$

in which FL is the set of followers of node i.

4.2 Signed Cascading Process

To compute the degree membership of each node to the leaders, we use a cascading process named network coordination game. In network coordination games all nodes have the same behaviour, but at a certain point in time, a set of nodes changes the behaviour. We are interested to know which and how many nodes are affected by this change. In other words, if all the nodes in the network have behaviour A and at some point in time a set of nodes changes the behaviour to B, then we are interested to know which nodes will be affected by behaviour B. In signed social networks, this process is a little different. For instance, let us consider that node i has behaviour B and node j and k has behaviour A. Although, the link between i and j is negative and the link between i and k is positive, j will not receive any pay-off from node i and even resists with behaviour A. On the contrary, k receives some pay-off and prefers to change its behaviour to the behaviour of B. One should compute the received pay-off by a node in order to check whether the node changes its behaviour. Therefore, node i's pay-off for a new behaviour A can be computed as follows:

$$Pay - Off(i) = \frac{\#(j_A \in Nei^+(i)) - \#(j_A \in Nei^-(i))}{\#(j_A \in Nei^+(i)) + \#(j_A \in Nei^-(i))}. \tag{3.7}$$

In the above formula j_A has behaviour A which is different than the current behaviour of node i. If the received pay-off by node i is bigger than the node i's threshold, then the node changes its behaviour. In the experiments, we consider local leadership computed in formula 3 as the threshold value of each node. We equal the number of leaders to the number of communities. To compute degree membership of a node to a certain community j, we change the behaviour of the community j's leader. In other words, other nodes in the network are affected based on their dependence on the corresponding leader of community j. Membership of each node to the community is computed based on the time iteration in which nodes change their behaviour. For node i, membership to community C_j is computed based on the following:

$$M_{ij} = \frac{1}{t_{ij}^2}. \tag{3.8}$$

where t_{ij} is the time iteration in which node i changes its behaviour to the behaviour of the community j's leader. M_{ij} is the soft membership value of node i belonging to community j.

4.3 Overlapping Community-Based Ranking Algorithms

Now with the help of two-phase OCD algorithm described in Sect. 4, we are able to identify overlapping community structures in signed networks. As we need to investigate effect of overlapping nodes, we proceed by using the community information and devising a overlapping community-aware ranking algorithm. In fact, HITS and PageRank can help to connect OCD and link prediction.

4.3.1 Overlapping Community-Based HITS

Firstly we change updating rules of HITS. It works based on two basic vectors named hubs and authorities. First, these two vectors are initialized with some random values, then they are updated until convergence. However, they merely take into account incoming and outgoing connections and do not consider overlapping community dimension. In order to consider overlapping community structures, we propose an extended version of the HITS algorithm. We not only differentiate between intra and extra community links but also take into account overlapping community structures. As there exist both positive and negative connections, we consider four types of vectors as follows:

$$a^+_i = \alpha \times \sum_{j \in indeg^+_{intra}(i)} h^+_j + \beta \times \sum_{j \in indeg^+_{ovl}(i)} h^+_j + \gamma \times \sum_{j \in indeg^+_{extra}(i)} h^+_j$$

$$a^-_i = \alpha \times \sum_{j \in indeg^-_{intra}(i)} h^-_j + \beta \times \sum_{j \in indeg^-_{ovl}(i)} h^-_j + \gamma \times \sum_{j \in indeg^-_{extra}(i)} h^-_j$$

$$h^+_i = \alpha \times \sum_{j \in outdeg^+_{intra}(i)} a^+_j + \beta \times \sum_{j \in outdeg^+_{ovl}(i)} a^+_j + \gamma \times \sum_{j \in outdeg^+_{extra}(i)} a^+_j$$

$$h^-_i = \alpha \times \sum_{j \in outdeg^-_{intra}(i)} a^-_j + \beta \times \sum_{j \in outdeg^-_{ovl}(i)} a^-_j + \gamma \times \sum_{j \in outdeg^-_{extra}(i)} a^-_j$$

(3.9)

in which a^+_i and a^-_i are, respectively, positive and negative authorities of node i, h^+_i and h^-_i are positive and negative hubs of node i. Original HITS algorithm considers random values for hubs and authorities initialization. It is proved that the algorithm converges after sufficient number of iterations [20]. We initialize hubs and authorities with $\frac{1}{\#V}$ and runs the updating rules for sufficient number of iterations for OC-HITS. Although we were not able to provide convergence proof for our algorithm, hubs and authorities are updated sufficiently. Finally, α, β and γ, respectively, contribute to *intra*, *overlapping* and *extra* hub and authority nodes.

4.3.2 Overlapping Community-Based PageRank

In this section we extend classical PageRank algorithm to also consider overlapping members in social media. For each node we consider both positive and negative PageRank value. Positive and negative overlapping community-based updating rules can be well described as follows:

$$PR^+_i = \zeta \times (\alpha \times \sum_{j \in indeg^+_{intra}(i)} (\frac{PR^+_j}{\#Outdeg^+_{intra(j)}}) + \beta \times \sum_{j \in indeg^+_{ovl}(i)} (\frac{PR^+_j}{\#Outdeg^+_{ovl(j)}}) +$$

$$\gamma \times \sum_{j \in indeg^+_{extra}(i)} (\frac{PR^+_j}{\#Outdeg^+_{extra(j)}})) + (1 - \zeta) * (\frac{1}{\#V})$$

$$PR^-_i = \zeta \times (\alpha \times \sum_{j \in indeg^-_{intra}(i)} (\frac{PR^-_j}{\#Outdeg^-_{intra(j)}}) + \beta \times \sum_{j \in indeg^-_{ovl}(i)} (\frac{PR^-_j}{\#Outdeg^-_{ovl(j)}}) +$$

$$\gamma \times \sum_{j \in indeg^-_{extra}(i)} (\frac{PR^-_j}{\#Outdeg^-_{extra(j)}})) + (1 - \zeta) * (\frac{1}{\#V})$$

(3.10)

where PR^+_i and PR^-_i are positive and negative PageRank values of node i, ζ is damping factor in the original PageRank set to 0.85 and α, β and γ coefficients weigh to intra, overlapping and extra nodes linking to node i.

4.4   Baseline OCD Methods

To evaluate SDMID, we compare it with two other community detection algorithms in signed networks named SPM and MEAs-SN. These algorithms are described in the following.

4.4.1   Signed Probabilistic Mixture Model

The SPM model is a variant of the probabilistic mixture model which produces edges with certain probabilities [8]. It is based on expectation-maximization (EM) method, which tries to maximize the probability containing latent variables. In this model, the number of communities (K) has to be defined beforehand. ω_{rs} is the probability of an edge e_{ij} choosing a community pair $\{r, s\}$ ($1 \leq r, s \leq k$) with the constraint $\sum_{rs} \omega_{rs} = 1$. e_{ij} is located in one community if $r=s$ and is between two communities if not. The probability of community r (s) choosing node i (j) is denoted as θ_{ri} (θ_{sj}). For any community r, given n nodes in the network, $\sum_i \theta_{ri} = 1$. As a result, the edge probability by SPM is as follows:

$$P(e_{ij}|\omega, \theta) = \left(\sum_{rr} \omega_{rr} \theta_{ri} \theta_{rj} \right)^{A_{ij}^+} \left(\sum_{rs(r \neq s)} \omega_{rs} \theta_{ri} \theta_{sj} \right)^{A_{ij}^-}. \qquad (3.11)$$

Here the chosen community pair of the edge in the above equation is a latent variable. We update θ_{ri} and θ_{sj} by minimizing $P(e_{ij}|\omega, \theta)$. This algorithm provides a soft partition of the concerned network, as a node can belong to several communities simultaneously [8].

4.4.2   Multi-Objective Evolutionary Algorithm in Signed Networks

This evolutionary algorithm is based upon the so-called structural similarity between adjacent nodes [28]. The similarity of two nodes in a signed network is defined as:

$$s(u, v) = \frac{\sum_{x \in B(u) \cap B(v)} \psi(x)}{\sqrt{\sum_{x \in B(u)} w_{ux}^2} \cdot \sqrt{\sum_{x \in B(v)} w_{vx}^2}}. \qquad (3.12)$$

where $B(u)$ ($B(v)$) is the set of node $u(v)$ and u's(v's) neighbours and $w_{ux}(w_{vx})$ is the weight of the edge connecting $u(v)$ and x. Besides, two objective functions are designed and employed to maximize the sum of positive similarities within communities and the sum of negative similarities between communities. Optimal solution is achieved by using a multi-objective evolutionary algorithm based on decomposition [28].

5   Sign Prediction

We are interested to investigate the effect of overlapping nodes in the sign prediction problem. To this end, we apply different classifiers to avoid possible biases imposed by using a classification approach. In fact, we show that overlapping nodes are significant, even by employing different classification methods. In the following subsections, classifiers and feature sets are explained.

5.1   *Classifiers*

In this section, we introduce classifiers which are used in the experiments.

5.1.1   Logistic Regression

Logistic regression uses a set of features for training and makes a probabilistic model out of the training data. The logistic regression utilizes a cost function to find the parameters of the model. The cost function is as follows:

$$J(h_\theta(x), y) = \frac{1}{2} \times \left(\frac{1}{1 + \exp(-\theta^T x)} - y \right)^2 . \qquad (3.13)$$

where J is the error in the model, x is the vector of features, θ is the vector of parameters which should be computed by the optimization approach. Finally, y contains the real class values. In our case, it is a vector of $+1$ and -1 values [5].

5.1.2   Bagging

The bagging classifier uses a number of weak classifiers in which their performance is better than a random classifier. Each of them is trained with some portion of the training data. Finally, the predicted class is decided based on the aggregated votes of the whole classifiers. This aggregation decision process is known as bagging or bootstrap aggregation [5].

5.1.3   J48

J48 is another classifier which is categorized in the family of decision trees and uses features to constitute decision tree over the training data [29]. In sign prediction, leafs of the tree would be -1 and $+1$ nodes.

5.1.4   Decision Table

Decision tables infer some rules for predicting the classes. Flowcharts and if-then-else statements are created to deduce the closest classes.

5.1.5   Bayesian Network and Naive Bayesian

Bayesian Network (BayesNet) and Naive Bayesian (BayesNaive) are based on the Bayesian probability theory. Bayesian Network is a form of probabilistic graphical models which uses directed graphs to show probability distributions. Each node in BayesNet indicates a random variable and edges denote the transition probability to transfer from one state to the other. It uses training data to build the model and computation of the probability dependencies. Based on statistical inference, it classifies test cases to their predicted classes. Additionally, BayesNaive applies the original Bayesian theorem to make the statistical inference of the classes which in our case are -1 and $+1$ values [5].

5.2   Sign Prediction Features

5.2.1   Simple Degree Sign Prediction Features

Link prediction approaches have applied degree features to prediction tasks such as link or sign prediction. Leskovec et al. employed simple degree features such as $indeg^+$, $indeg^-$, $outdeg^+$, $outdeg^-$ and number of common neighbours between trustor and trustee as for the features [25]. Moreover, status and balancing theory in signed networks have been applied to the case of sign prediction. Other proximity measures such as jaccard index and embeddedness are also applied to predict proximity of users in social networks [24]. Similarly for the features, we employ simple in-degree and out-degree features which correspond to prestige and opinions of nodes. However, we delicately change them to examine the effect of overlapping, intra and extra nodes. In Table 3.1, one can observe three sets of features which are applied for training and test purposes.

Table 3.1 Three sets of features including intra, overlapping and extra features which have been used for the classifier

Intra	Ovl	Extra
$\#indeg_{intra}(u)$	$\#indeg_{Ovl}(u)$	$\#indeg_{extra}(u)$
$\#outdeg_{intra}(u)$	$\#outdeg_{Ovl}(u)$	$\#outdeg_{extra}(u)$
$\#indeg_{intra}(v)$	$\#indeg_{Ovl}(v)$	$\#indeg_{extra}(v)$
$\#outdeg_{intra}(v)$	$\#indeg_{Ovl}(v)$	$\#indeg_{extra}(v)$

Here u and v, respectively, refer to trustor and trustee. *All* set is not listed and it comprises all the 12 features

To compute in-degree feature types for a node i we use the following formula:

$$indeg_{type}(i) = \frac{\#indeg^+_{type}(i) - \#indeg^-_{type}(i)}{\#indeg^+_{type}(i) + \#indeg^-_{type}(i)}, \tag{3.14}$$

where $type \in \{Intra, Extra, Ovl\}$ and i can be either u or v. Respectively, $outdeg_{type}(i)$ feature types can be computed as follows:

$$outdeg_{type}(i) = \frac{\#outdeg^+_{type}(i) - \#outdeg^-_{type}(i)}{\#outdeg^+_{type}(i) + \#outdeg^-_{type}(i)}. \tag{3.15}$$

Therefore, for each edge corresponding to a trustor and trustee, one can consider these three sets of features. In the *Intra* type, nodes which are in the same community as trustor and trustee, are considered and in-degree and out-degree features of this type are computed based on these nodes. Similarly, overlapping type comprises nodes in which are overlapping among communities and are neighbours to trustor and trustee. Finally, *Extra* type features consider nodes which are in different communities than trustor and trustee but are neighbours to them.

5.2.2   OC-HITS Sign Prediction

To employ OC-HITS for the case of sign prediction, we apply final positive and negative hubs and authority vectors to the case of sign prediction. In other words, for each combination of α, β and γ values, final a^+_u, a^-_u, h^+_u and h^-_u of trustor and trustee are considered as features. Each set of α, β and γ values weigh to opinions of intra, overlapping and extra nodes. For instance, $\alpha = 0.3$, $\beta = 0.4$ and $\gamma = 0.3$ mean that overlapping nodes are considered more.

5.2.3   OC-PageRank Sign Prediction

Similarly we apply OC-PageRank to the case of sign prediction, we simply use positive and negative node PageRank values. In other words, PR^+_u and PR^-_u of both trustor and trustee are employed to build the prediction model. Similar to HITS, for each combination of α, β and γ values, a prediction model is built and MAE, RMSE and prediction accuracy value is computed from test set.

6   Dataset and Metrics

In this section, we introduce the datasets[1] and evaluation metrics which are used in the experiments.

6.1   Real World Networks

Wikipedia Election (Wiki-Elec): Wikipedia is a free glossary which is collaboratively written by users from all around the world. There exist different users with different levels of authorization. Administrators have higher access level for editing and other actions. Also, there exist elections for selecting these administrators. Wikipedia users can positively and negatively vote toward the candidates. Hence, it is a signed social network.

Wikipedia Request for Adminship (Wiki-RfA): This is an updated version of the Wikipedia adminship election data. This dataset is enriched with textual information and the result of the voting for the adminship election. Similarly, users vote positively and negatively toward the candidates. One can observe more information about these two datasets in Table 3.2.

6.2   Synthetic Networks

Because of the unavailability of real world datasets with respective ground truth for evaluating OCD algorithms in signed networks, a generator which produces signed graphs with overlapping community structures and ground truth information is required. Liu et al. have developed a synthetic network generator, which combines the LFR model [23] and a model proposed by Yang et al. [50]. Referring to the idea of [28], synthetic networks will be developed based on the directed unweighted LFR model.[2] The parameters of the LFR model include the number of nodes n, the average and maximum degree of each node k and *maxk*, the minus exponents

Table 3.2 Number of nodes, edges, positive and negative links related to Wiki-RfA and Wiki-Elec

	#Nodes	#Edges	#PositiveEdges	#NegativeEdges
Wiki-RfA	10,835	159,388	144,451	41,176
Wiki-Elec	8298	103,186	83,962	23,118

---

[1] https://snap.stanford.edu/data/

[2] https://sites.google.com/site/santofortunato/inthepress2

for the degree and community size distributions following power laws t_1 and t_2, minimum and maximum community size *minc* and *maxc*, the number of nodes in overlapping communities *on*, the number of communities they belong to *om* and the fraction of edges that each node shares with the other nodes outside of its community μ. An LFR network is then adapted to produce a signed synthetic network by using two more parameters: the fractions of negative connections within communities P_- and positive connections between communities P_+, which adjust the noise level of the concerned synthetic network. A generated LFR network is adapted by using two more parameters: the fractions of negative connections within communities P_- and positive connections between communities P_+, which adjust the noise level of the concerned synthetic network [28]. Concretely speaking, the following procedure is carried out on a generated LFR network: weights of all edges between communities are negated, then randomly selected P_- fractions of edges inside communities are negated and afterwards randomly selected P_+ fractions of edges between communities are negated.

6.3   Evaluation Metrics

6.3.1   Normalized Mutual Information

The knowledge-driven metrics such as normalized mutual information (NMI) is applied to evaluate community detection algorithms, because it focuses on the question whether a node resides in the right community (communities) without considering issues like node degree or edge weight [23]. As a result, the prerequisite of using this metric is the known ground truth. Originating from the information theory, the NMI regards the k-th entry of a node i's membership vector M_{ik} in the detected community structure and the vector $M_{il'}$ in the ground truth as two random variables and determines the mutual information shared by these two variables. The normalized aggregate of membership vectors of all nodes corresponds to the extended NMI result. It ranges from 0 to 1 and a big value suggests a high similarity.

6.3.2   Modularity

The statistic metric modularity is applied to evaluate community detection algorithms in our paper [13]. Modularity was originally proposed by Newmann and Girvan [32]. It takes the degree distribution of nodes into consideration and is a widely applied measure to evaluate the quality of dividing a network into modules or communities. The basic idea is that the number of edges falling into a community should be higher than the expected number of edges between these nodes, if the edges between the nodes had been randomly shuffled. Gomez et al. later extended modularity to address signed networks with disjoint communities. We use the modularity in signed networks to compare the algorithms [13].

Fig. 3.1 In signed graphs, triangles can form 4 states which 2 of them are balanced and the other 2 are unbalanced and have tendency to become balanced in some point in time

6.3.3 Frustration

Frustration can be computed for crisp community detection algorithms in signed social networks. Triangles have four states of balanced and unbalanced in networks with positive and negative links. As Fig. 3.1 indicates, two of the states are balanced and the other two are unbalanced [6]. It can be proven that a signed network is balanced when nodes inside communities have positive relationships and nodes among clusters have negative relationships with each other. The quality of a community detection algorithm can be estimated by the number of edges that do not satisfy the balancing property. Hence, we can compute the frustration of identified communities for $\forall i, j \in C$ as follows:

$$Frustration = \alpha \times \#C_j^- + (1 - \alpha) \times \#C_{ij}^+, \tag{3.16}$$

where $\#C_j^-$ is the number of edges in community j that breaks the balancing theory. Moreover, $\#C_{ij}^+$ is the number of positive edges among communities of i and j that breaks the theory of balance.

7 Results

In this section, experimented results are demonstrated in detail.

7.1 Results of OCD

The experiments on benchmark networks changed one single parameter at each time. A network size $n = 100$ nodes was applied, so that it would not take too long to run the experiments. Unless specified, the other default parameter values were set as follows: $k = 3$, $maxk = 6$, $\mu = 0.1$, $t_1 = -2.0$, $t_2 = -1.0$, $minc = 5$, $maxc = 30$, $on = 5$, $om = 2$, $P_- = 0.1$, $P_+ = 0.1$. Every experiment was carried out three times and each metric value would take their average. SPM requires the prior knowledge about the total number of communities, which is unrealistic; without community detection the community number can hardly be known. Thus, for all experiments, SPM is regarded as a bisection algorithm. In addition, the number of trials of EM can be parametrized for SPM. In order to avoid long running time but ensure a better result, the number of trials of EM was set to 3 considering the small size of the network. With the focus on strongly and moderately attached members, the iteration boundary of the network coordination game was set to 3 for SDMID. In these subsections, unless otherwise specified, the x-axis of figures always indicates the respective parameter whose value has been changed in order to see the effect on algorithm performance while the y-axis represents the accordant metric value. In all figures, for the sake of simplicity, MEA_s-SN is denoted as MEA.

7.1.1 Network Size n

The network sizes were set between 50 and 400. It can be seen from Fig. 3.2 that SDMID were much faster than the other two with execution time ranging from 5 to 39. The execution time of MEA_s-SN followed an exponential growth while that of SPM were rather volatile. Due to the fact that the maximum value along the y-axis in the execution time graph is very large, the execution time of SDMID cannot be easily read. In Table 3.3, the exact values can be found. With the increasing network size, SDMID demonstrated a stable short execution time, leaving the other two algorithms far behind. For example, with a network size of 400, SDMID's execution time was 18 ms, which was $\frac{1}{1742}$ of SPM's and $\frac{1}{10289}$ of MEA_s-SN's. SDMID had the highest modularity in most cases while it was caught up by SPM when the network size increased. SPM had a clear advantage over the other two algorithms in terms of frustration. SPM also enjoyed higher NMI values when the network size was small ($n = 50$ and $n = 100$). However, its NMI fell exponentially with the increasing size of the network. Presumably because of the limitation of maximum community

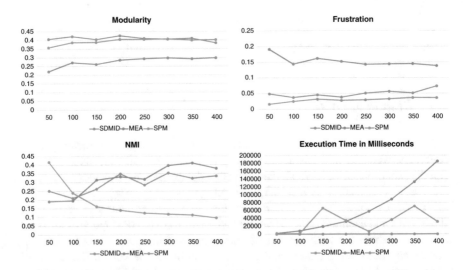

Fig. 3.2 Performance overview with different network size

Table 3.3 Execution time with different network size

	Execution time in millisecond		
Network size	SDMID	MEA	SPM
50	37	1915	1037
100	19	8129	1377
150	16	19,258	66,235
200	20	31,816	34,573
250	18	57,849	6720
300	22	88,954	36,575
350	21	133,096	70,973
400	18	185,197	31,348

Parameters: n = 100, k = 3, maxk = 6, μ = 0.1, t1 = −2.0, t2 = −1.0, minc = 5, maxc = 30, on = 5, om = 2, P_ = 0.01, P_+ = 0.01

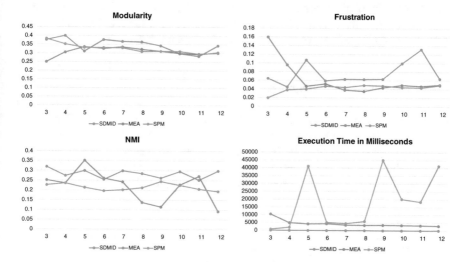

Fig. 3.3 Performance overview with different average node degree

size to 30, the generated synthetic network of a larger size tended to have numerous communities, which SPM with predefined community number as 2 could not catch up. On the contrary, SDMID and MEA$_s$-SN produced better NMI value where n was fairly large: SDMID had its highest value (0.354) at $n = 300$ while MEA$_s$-SN had its highest value (0.411) at $n = 350$.

7.1.2 Average Node Degree k

In this experiment, the average degree varied between 3 and 12. The maximum degree was set to 15, so that synthetic networks could be produced even with a high average degree. Figure 3.3 shows that SDMID still had the highest modularity in most cases. With $k \in \{3, 5, 10, 11\}$, SDMID lagged slightly behind, with the respective differences to the best values as 0.006, 0.025, 0.012 and 0.013. However, with the increasing average degree its modularity value did tend to decrease. Presumably because SDMID is a leader-based algorithm, it is good in handling networks featured with leader–follower relations, where leaders have far more edges than their followers. With the increasing average node degree, the difference between leaders and their followers in terms of degrees became smaller. SDMID still took a leading place regarding execution time. SPM had a volatile execution time and tended to require more run time with the increasing number of k, which indicated an increasing number of edges.

SDMID had also the highest NMI value in 70% cases. The NMI value of MEA_s-SN moved within a wide range: when $k = 5$, it achieved the best value of the three algorithms, which was equal to 0.352; when $k = 12$, it reached the lowest value of 0.093 throughout this experiment. SPM enjoyed the best frustration value in most cases, i.e., 7 out of 10. It was beaten by MEA_s-SN with $k \in \{7, 8, 9\}$. SPM, on the other hand, displayed the worst frustration value when $k \geq 5$.

7.1.3 Maximum Node Degree *maxk*

In this experiment the maximum node degree varied between 5 and 16. Seen from Fig. 3.4, SDMID was the best regarding execution time. SPM took the second place in terms of execution time in 10 out of 12 cases. However, it took SPM very long to complete the cover detection when *maxk* = 6 and 16. Especially when *maxk* = 16, the execution time soared up to 48,040 ms, which was about 9600 times

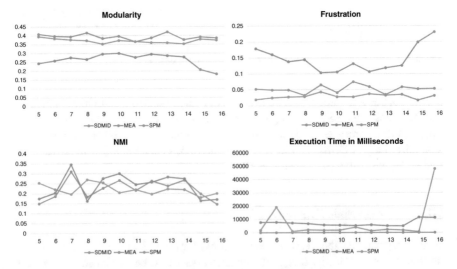

Fig. 3.4 Performance overview with different maximum node degree

of SDMID's and 4 times of MEA_s-SN's value. SDMID also did better than the other two algorithms in terms of modularity in all cases but one. It exerted but only slight advantage over SPM. When $maxk = 11$, SPM beat SDMID with a value of 0.366 against 0.365. SPM had the lowest frustration value in all cases, but was almost caught up by SDMID when $maxk$ was 8 and 13. MEA_s-SN had a weak advantage over its peers in terms of NMI value: in 50% cases, it secured the top value.

7.1.4   Fraction of Edges Sharing with Other Communities μ

In this experiment, the fraction of edges of each node, which are connected to nodes in other communities, changed from 0.05 to 0.185, with the step of 0.05. A higher μ value indicates a higher interaction degree between communities, which makes it more difficult to detect communities. As a matter of fact, Fig. 3.8 shows that the modularity of all algorithms demonstrated a downward trend. SDMID had its highest value of 0.431 when $\mu = 0.08$ and its lowest value of 0.350 when $\mu = 0.185$. The trend was more obvious with SPM and MEA_s-SN: the value of SPM decreased from 0.413 to 0.331 while that of MEA_s-SN decreased from 0.289 to 0.230.

SPM still showed the best frustration value. At the same time, frustration of SPM displayed an upward trend: SPM had its best value equal to 0.014 when $\mu = 0.05$ and its worst value 0.047 when $\mu = 0.155$. SDMID had its best frustration value equal to 0.025 when $\mu = 0.08$, which was even slightly lower than that of SPM. Within the range $\mu \in [0.014, 0.185]$, frustration of SDMID also displayed an upward trend: it increased from 0.040 to 0.062. MEA_s-SN did the best in terms of NMI.

7.1.5   Maximum Community Size maxc

In this experiment, the maximum community size ranged from 30 to 40. It can be seen from Fig. 3.5 that SDMID still had the shortest execution time, followed by SPM. MEA_s-SN lagged behind with even big disparity to SPM. For example, when $maxc = 40$, it took SDMID 9 ms, SPM 2080 ms and MEA_s-SN 10,096 ms to detect their respective covers. SDMID also demonstrated the highest modularity value with only one exception. At $maxc = 33$, SPM's value lied a little higher than SDMID's, with 0.380 against 0.379. Modularity value of all algorithms moved only within a small corridor: $[0.379, 0.425]$ for SDMID, $[0.243, 0.317]$ for MEA_s-SN and $[0.366, 0.383]$ for SPM. SPM was leading in terms of frustration and SDMID followed closely. The average difference between these two algorithms is only 0.015. The frustration values of MEA_s-SN were much higher than those of its peers, lying between 0.099 and 0.176.

Regarding NMI, SPM displayed a clear upward trend till $maxk = 39$ with its value rising from 0.194 to 0.258. When $maxk$ was between 34 and 40, the NMI value of MEA_s-SN demonstrated a slight upward trend as well, rising from 0.216 to

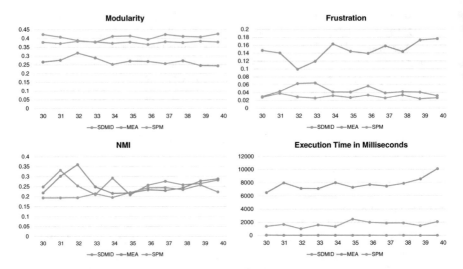

Fig. 3.5 Performance overview with different maximum community size

0.288. When *maxk* was greater than 34, the NMI value of SDMID kept increasing with the exception at *maxk* = 37, rising from 0.217 to 0.288. The reason might be that a greater maximum size of communities indicates a clear community structure and therefore ease the community detection for all algorithms.

7.1.6 Number of Nodes in Overlapping Communities *on*

In this experiment, the *on* value varied from 1 to 10. Seen from Fig. 3.6, SDMID still had the best value for execution time. In 90% cases, it also had the best modularity value. Only when *on* = 1, it had a lower value than SPM, with 0.352 against 0.378. It can also be observed that SDMID improved in modularity as a whole. Besides, its advantage over the other two algorithms became stronger as the *on* increased. With *on* = 3, the modularity value of SDMID was 0.013 higher than SPM and 0.09 higher than MEA_s-SN; while *on* = 10, it was 0.04 higher than SPM and 0.13 higher than MEA_s-SN. SDMID also improved in frustration when *on* increased. When *on* = 1, its frustration value lied far higher than SPM, with 0.092 against 0.025. When *on* = 9, its value was almost as low as that of SPM, with 0.031 against 0.030. MEA_s-SN had the best NMI value in 5 out of 10 cases. However, this advantage was rather weak when *on* was 5 and 7: the differences to the second best values were both about 0.01. SDMID demonstrated the highest NMI value when *on* ≥ 7. Altogether, SDMID improved relatively to its peers in modularity, frustration and NMI. Presumably because the network coordination game employed in SDMID can detect nodes residing in different communities even with weak attachment.

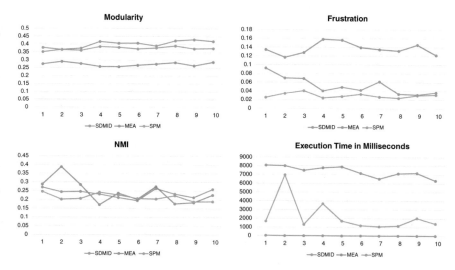

Fig. 3.6 Performance overview with different number of nodes in overlapping communities

7.1.7   Number of Communities Which Nodes in Overlapping Communities Belong to *om*

In this experiment, *om* ranged between 2 and 10. The network size was set to 200 so that networks even with the highest number of overlapping communities could be generated. As a result, the number of *on* was also proportionally raised to 10. As shown in Fig. 3.7, SDMID was significantly better than the other two algorithms in execution time. SPM followed as the second in most cases with one exception at *om* = 4, where its execution time rose to 51,461, which was 2617 times of SDMID's value and 1.5 times of MEA_s-SN's value. SDMID was also superior to the other two algorithms in modularity in 90% cases, with only one exception at *om* = 10, where it was beaten by SPM (SDMID: 0.392, SPM:0.393). SPM had a slight advantage over SDMID in terms of frustration with one exception. At *om* = 6, SDMID had a value of 0.0321 while SPM had a value of 0.0323.

As for NMI, MEA_s-SN took the lead when *om* was between 2 and 4. When *om* > 4, SDMID caught up and its value exceeded that of MEA_s-SN. Only at *om* = 8 the NMI value of SDMID was a slightly lower than that of MEA_s-SN, with 0.203 against 0.205. It can also be observed that the NMI value of SPM was somewhat far away from that of the other two algorithms and it decreased as *om* became bigger. Obviously, the predefined community number as two limited the possibility of SPM's cover conforming to the real cover (Fig. 3.8).

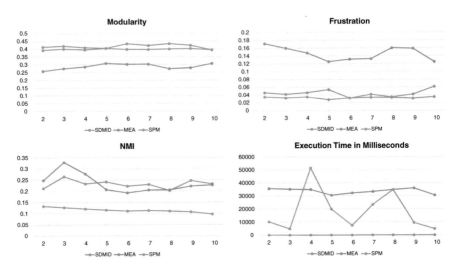

Fig. 3.7 Performance overview with different number of communities which nodes in overlapping communities belong to

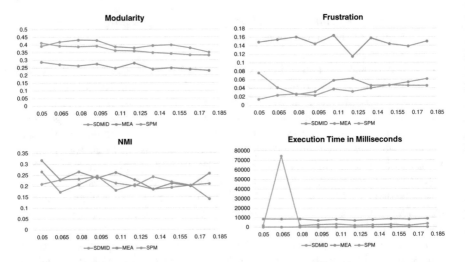

Fig. 3.8 Performance overview with different fraction of edges sharing with other communities

7.1.8 Fractions of Positive Connections Between Communities P_+

A higher value of P_- or P_+, similar to μ, suggests that the concerned community structure is more ambiguous while a lower one indicates a clearer structure. In this experiment, the P_+ value ranged from 0.01 to 0.1.

Seen from Fig. 3.9, SDMID exerted an absolute advantage in execution time, with 5 and 46 ms as the lowest and highest value, respectively. The value of MEA_s-SN moved between 6388 and 9635 ms while that of SPM moved between 1477

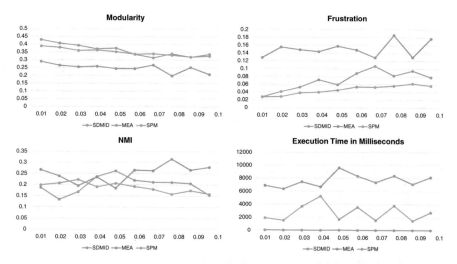

Fig. 3.9 Performance overview with different fractions of positive connections between communities

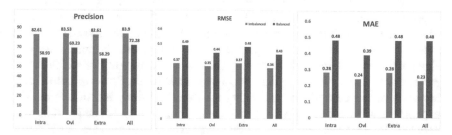

Fig. 3.10 Prediction accuracy, RMSE and MAE results on the Wiki-Elec dataset. Intra, overlapping (Ovl) and all of them together (all) are used for the sign prediction task. It can be observed that in both balanced and imbalanced cases, *Ovl* feature set has better performance in comparison to *Intra* and *Extra*. This holds for all the metrics including prediction accuracy, RMSE and MAE

and 5286 ms. SDMID also produced the highest modularity values among the three algorithms in most cases, followed closely by SPM. At $P_+ = 0.07$ and $P_+ = 0.09$, the value of SPM was 0.025 and 0.0003 higher than that of SDMID. SPM had the best frustration value and SDMID took the second place. MEA$_s$-SN did best in terms of NMI: in 7 out of 10 cases, it had the highest value. It is also shown that modularity and frustration values of all algorithms got worse with the increasing P_+, presumably resulting from the greater ambiguity of the community structure (Figs. 3.10 and 3.11).

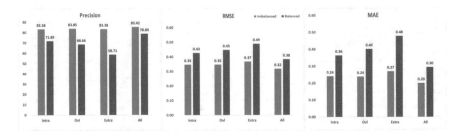

Fig. 3.11 Prediction accuracy, RMSE and MAE results on Wiki-RfA dataset. Intra, overlapping (Ovl) and all of them together (all) are used for the sign prediction task. It can be observed that the performance of *Ovl* feature set is higher than *Intra* and *Extra* feature sets. Only for balanced case *Intra* is a little bit better than *Ovl*. However, both *Intra* and *Ovl* feature sets performed better than *Extra* feature set. This also holds for all the metrics including prediction accuracy, RMSE and MAE

7.1.9 Experiments on Real World Network

The three algorithms have also been run on wiki-Elec.[3] In the following, the experiment result on wiki-Elec will be analysed. For SPM, a predefined number of communities as 2 did not make much sense because of the large size of this network. So it was set to 5 in the experiment, which [16] finds to be optimal to make sign predictions in large scale networks. The number of trials of EM which can be parametrized was set to 3 in order to achieve a ideal result. However, considering the scale of the network, in calculating the execution time, the value from the experiment was divided by 3 for a fair comparison. As the execution time was very long for SPM and MEA_s-SN, its value was expressed in minutes. For SDMID, the network coordination game was again set to 3 iterations, leaving out loosely connected nodes for each community. From Fig. 3.12, it can be seen that SDMID had an absolute advantage over the other two algorithms in execution time: with 0.16 min, its value accounted only for $\frac{1}{11000}$ of SPM's and $\frac{1}{19381}$ of MEA_s-SN's. SDMID was also superior to the other two algorithms in terms of modularity: its value was 1.09 times of SPM's and 1.36 times of MEA_s-SN's. The three algorithms produced similar frustration results: 0.0953 for SPM, 0.0962 for SDMID, 0.1089 for MEA_s-SN.

7.2 Simple Degree Sign Prediction Results

After identifying communities and computing intra, extra and overlapping (ovl) feature types, they are applied to the case of sign prediction problem. To this end, we use the WEKA software and recruit bagging, BayesNaive, BayesNet, J48, logistic

[3]https://snap.stanford.edu/data/wiki-Elec.html

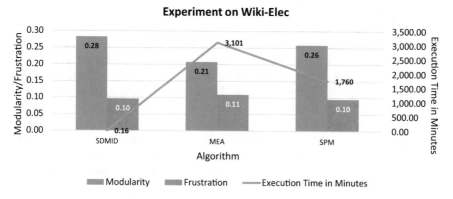

Fig. 3.12 Performance result of experiment on Wiki-Elec

regression and decision table as classifiers. Tenfold cross validation is used for all the algorithms. Moreover, to evaluate the results of models, we use prediction accuracy, Mean Average Error (MAE), Root Mean Square Error (RMSE) [45]. Because the number of positive edges is more than the number of negative ones, we consider both balanced and imbalanced datasets. On the one hand, imbalance refers to the original dataset in which the number of positive and negative edges is unequal. On the other hand, the balance dataset is a subset of the imbalanced dataset in which the number of positive and negative links is randomly equalized. In order to prevent errors, the results for the balanced dataset are performed 10 times with random omitting of positive edges. Then, the results are averaged over these 10 experiments. Before train and test phase, we preprocessed the data and omitted those edges in which their feature sets are completely empty.

Figures 3.10 and 3.11 indicate results of the logistic regression classier on Wiki-Elec and Wiki-RfA, respectively. As for Wiki-Elec in imbalanced case, prediction accuracy values are 82.61, 83.53, 82.61 and 83.9, respectively, for *Intra*, *Ovl*, *Extra* and *All* sets of features. In this case, *Ovl* feature set has roughly 1 % better performance in comparison to *Intra* and *Extra* features. One could also observe the same pattern in MAE and RMSE. RMSE error for *Ovl* feature type set is 0.35 which is lower than *Intra* (0.37) and *Extra* (0.37) feature sets. Furthermore, MAE error for *Ovl* feature set is 0.24 which is lower than *Intra* (0.27) and *Extra* (0.27) feature sets. As for Wiki-Elec in balanced case, the difference is more tangible regarding these measures. As one can observe, prediction accuracy for *Ovl* features is 69.23 which is approximately 10 % higher than *Intra* (58.93) and *Extra* (58.29) features. This also holds for MAE and RMSE. MAE value for *Ovl* (0.39) is lower than *Intra* (0.48) and *Extra* (0.48). Finally, *Ovl* (0.44) surpasses *intra* (0.49) and *Extra* (0.48) in terms of RMSE.

Similarly, one can observe approximately the same pattern for the Wiki-RfA. As Fig. 3.11 indicates for the imbalanced case, prediction accuracy of Ovl (83.85) supersedes by 0.5 % over the *Intra* (83.38) and *Extra* (83.38) feature sets. Interestingly, in balanced dataset the pattern is a little bit different from what we observed in Wiki-Elec and the imbalanced case of Wiki-RfA. Here, the *Intra* feature set obtains prediction accuracy of 71.89 which is higher than the *ovl* (68.66) and *Extra* (58.71) feature sets. However, the *ovl* feature is about 10 % higher than the extra feature set which again approves the significance of overlapping set features. The same pattern can be observed for MAE and RMSE measures. In the imbalanced case, MAE value for the *Ovl* feature set (0.24) is equal to *Intra* (0.24) and lower than *Extra* (0.48). However, this is not true for the balanced case. *Intra* (0.36) is lower than both *Ovl* (0.40) and *Extra* (0.48) in terms of MAE. The same relation also holds for RMSE. In other words, *Ovl* (0.35) is equal to *Intra* (0.35) and lower than *Extra* (0.37) in the imbalanced case. However, *intra* (0.43) beats both *Ovl* (0.45) and *Extra* (0.49) in the balanced case.

Last but not least, we compared the performance of different classifiers on the feature sets of *Intra*, *Extra*, *Ovl* and *All* for imbalanced datasets. Results of the prediction accuracy comparison on Wiki-Elec and Wiki-RfA are, respectively, shown in Tables 3.4 and 3.5. In Wiki-Elec, one can observe for all the classifiers that the *Ovl* feature set is higher than the *Intra* and *Extra* feature sets except in BayesNaive classifier in which both *Intra* (82.61) and *Extra* (82.61) feature set prediction accuracy is higher than the *Ovl* set (80.37). Surprisingly, the *All* set has a lower performance. In terms of the performance of the classifier, we can observe for the *All* feature set, that Bagging (84.80), J48 (84.53), Decision Table (84.16) and logistic regression (83.90) have better performance in comparison to the other classifiers. Although, we applied logistic regression in experiments of Figs. 3.11 and 3.10, performance of Bagging, J48 and Decision Table are better.

Likewise, Table 3.5 indicates a similar pattern. For all the classifiers, prediction accuracy of the *Ovl* feature is higher than the *Intra* and *Extra* feature sets except in Bagging and BayesNaive. Furthermore, it can also be deduced that performance of Bagging (85.69), J48 (85.65), Logistic (85.43) and Decision Table (85.25) are higher than BayesNet (84.93) and BayesNaive (82.62) for the *All* feature set.

Table 3.4 Prediction accuracy of different classifiers on Wiki-Elec dataset

	Bagging	Logistic	BayesNet	BayesNaive	J48	Decision Table
Intra	82.61	82.61	82.61	82.61	82.61	82.61
Ovl	83.82	83.53	83.32	80.37	84.05	84.02
Extra	82.74	82.61	82.64	82.61	82.61	82.64
All	84.80	83.90	83.20	80.94	84.53	84.16

Each row of the table indicates the set of features which is used for training and test phase

Table 3.5 Prediction accuracy of different classifiers on Wiki-RfA dataset

	Bagging	Logistic	BayesNet	BayesNaive	J48	Decision Table
Intra	83.68	83.39	83.44	73.12	83.58	83.35
Ovl	83.67	83.85	83.99	82.92	83.88	83.93
Extra	83.51	83.39	83.4	83.39	83.39	83.42
All	85.69	85.43	84.93	82.62	85.65	85.25

Each row of the table indicates the set of features which is used for training and test phase

Table 3.6 MAE, RMSE and prediction accuracy values for different combination values of α (intra), β (Ovl) and γ (extra) coefficients in community-based HITS updating rules for Wiki-RfA dataset

α (intra)	β (Ovl)	γ (extra)	MAE	RMSE	Prediction accuracy
0.000	0.200	0.800	0.203	0.323	85.901
0.000	0.800	0.200	0.192	0.315	86.437
0.000	0.333	0.666	0.199	0.321	86.117
0.000	0.666	0.333	0.194	0.316	86.316
0.300	0.300	0.400	0.159	0.288	88.198
0.300	0.400	0.300	0.158	0.287	88.280
0.400	0.300	0.300	0.160	0.288	88.217
0.800	0.200	0.000	0.160	0.288	88.199
0.200	0.800	0.000	0.208	0.327	84.721
0.333	0.666	0.000	0.159	0.288	88.154
0.333	0.333	0.333	0.160	0.288	88.225
0.666	0.333	0.000	0.209	0.327	84.777
0.333	0.000	0.666	0.210	0.328	84.668
0.666	0.000	0.333	0.209	0.327	84.777
0.200	0.000	0.800	0.210	0.327	84.756
0.800	0.000	0.200	0.208	0.327	84.721

7.2.1 OC-HITS Sign Prediction

We employ bagging classifier to evaluate importance of intra, overlapping and extra nodes. For this purpose, 16 different combination of α (intra), β (overlapping) and γ (extra) values in Tables 3.6 and 3.7 are applied to the case of sign prediction. Table 3.6 indicates prediction accuracies, MAE and RMSE of prediction for different coefficient values in Wiki-RfA dataset. The lowest MAE (0.158), RMSE (0.287) and highest prediction accuracy (88.280) belong to $\alpha = 0.3$, $\beta = 0.4$ and $\gamma = 0.3$ which hubs and authority values of overlapping nodes are taken into account more than hubs and authorities of intra and extra nodes. In other cases like combination of 0.2, 0.8 and 0, one can observe that $(\alpha = 0.8, \beta = 0.2, \gamma = 0)$ leads to lowest MAE (0.16), RMSE (0.288) and highest prediction accuracy (88.199). Regarding combination of 0, 0.333 and 0.666, lowest MAE (0.159), RMSE (0.288) and highest prediction accuracy (88.154) relate to $(\alpha = 0.333, \beta = 0.666, \gamma = 0)$.

Table 3.7 MAE, RMSE and prediction accuracy values for different combination values of α (intra), β (Ovl) and γ (extra) coefficients in community-based PageRank updating rules for Wiki-RfA dataset

α (intra)	β (Ovl)	γ (Extra)	MAE	RMSE	Prediction accuracy
0.000	0.200	0.800	0.309	0.393	80.327
0.000	0.800	0.200	0.309	0.393	80.331
0.200	0.800	0.000	0.230	0.342	82.931
0.800	0.200	0.000	0.230	0.342	82.921
0.200	0.000	0.800	0.231	0.344	82.933
0.800	0.000	0.200	0.232	0.344	82.966
0.333	0.666	0.000	0.230	0.342	82.938
0.333	0.333	0.333	0.226	0.340	83.237
0.333	0.000	0.666	0.231	0.344	82.950
0.666	0.000	0.333	0.232	0.344	82.906
0.000	0.333	0.666	0.309	0.393	80.326
0.000	0.666	0.333	0.310	0.393	80.283
0.300	0.300	0.400	0.225	0.339	83.300
0.300	0.400	0.300	0.227	0.340	83.211
0.400	0.300	0.300	0.226	0.339	83.298
0.666	0.333	0.000	0.315	0.397	80.324

Regarding 0.3, 0.4 and 0.3, combination with higher coefficient for overlapping member wins ($\alpha = 0.3$, $\beta = 0.4$, $\gamma = 0.3$). Hence, one can observe that hub and authority values of overlapping members competitively affect sign prediction.

7.2.2 OC-PageRank Sign Prediction

Table 3.7 reveals MAE, RMSE and prediction accuracy of Wiki-RfA dataset using OC-PageRank. Similarly, one can observe that lowest MAEs, RMSEs and highest accuracies belong to $\alpha = 0.3$, $\beta = 0.3$, $\gamma = 0.4$ and $\alpha = 0.4$, $\beta = 0.3$, $\gamma = 0.3$ with, respectively, MAE values 0.225 and 0.226, RMSE values 0.339 and 0.339 and prediction accuracies of 83.300 and 83.298. For other combinations, high coefficient overlapping PageRank values are also competitive in Wiki-RfA. Other cases with high overlapping weight also possess competitive error values and prediction accuracies. Although prediction accuracy for OC-HITS is higher than OC-PageRank, the significant pattern of overlapping members is kept approximately the same. Moreover, comparing prediction accuracy results of simple *in and out*-degree, OC-HITS hub and authority and PageRank features, we observe that OC-HITS outperforms the others in Wiki-RfA.

8 Conclusion and Future Work

Social networks are popular platforms that have received incredible attention since more than a decade. The dynamic connections and overlapping structures within social media have raised open research questions. Among them we addressed "Are overlapping community structures significant in social networks?". To find out importance of overlapping structures and members, we needed an overlapping community detection and a way to connect community dimension to the sign prediction problem. Hence, we proposed a novel two-phase overlapping community detection for networks with positive and negative edges. Firstly, we identify leaders in the network based on the effective degree and dissimilarities among neighbours. Secondly, we computed membership of other nodes to leaders. We evaluated the proposed algorithm and the baselines on both synthetic and real world networks. We also applied overlapping nodes to the case of sign prediction problem. We extended original HITS and PageRank algorithms to overlapping community-based HITS (OC-HITS) and overlapping community-based PageRank (OC-PageRank) algorithms and investigated significance of overlapping members by employing the rank vectors to the case of sign prediction. We showed that the proposed overlapping community detection algorithm has good performance by having low frustration error, local dynamic and good accuracies when applied to the sign prediction problem. We indicated that overlapping nodes in comparison to other nodes in the network possess more information in terms of the structure and knowledge of the network.

We plan to run the proposed overlapping community detection algorithm on more signed social networks. Currently nature of datasets in signed networks are explicit votes of positive and negative connections. We intend to employ sentiment analysis techniques to forum data. Extracting positive and negative meaning from posts helps to identify implicit signed connections among users. Hence, we will be able to generalize our findings. Furthermore, we will investigate our methodology to improve sign prediction accuracies. Moreover, we are also interested to explore the effect of overlapping nodes in unsigned networks like learning environments and open source developer communities. Boundary spanners expand the borders of communities in learning environments and in open source developer communities. Therefore, we would like to investigate whether overlapping nodes show the same pattern as we observed it in this paper. Additionally, by observing importance of overlapping nodes, it worth employing the effect of overlapping community detection algorithms and overlapping nodes in devising new ranking and recommender algorithms. Finally regarding threads to validity, we need to compare effect of overlapping members to other social media with the help of other overlapping community detection algorithms as well as other applications, e.g., expert identification.

Acknowledgements The work has received funding from the European Commission's FP7 IP Learning Layers under grant agreement no 318209.

References

1. Amelio A, Pizzuti C (2013) Community mining in signed networks: a multiobjective approach. In: Proceedings of the 2013 IEEE/ACM international conference on advances in social networks analysis and mining (ASONAM '13). ACM, New York, NY, pp 95–99
2. Anchuri P, Ismail MM (2012) Communities and balance in signed networks: a spectral approach. In: 2012 IEEE/ACM international conference on advances in social networks analysis and mining (ASONAM), pp 235–242
3. Anchuri P, Magdon-Ismail M (2012) Communities and balance in signed networks: a spectral approach. In: Proceedings of the 2012 international conference on advances in social networks analysis and mining (ASONAM 2012). IEEE Computer Society, Washington, DC, pp 235–242
4. Backstrom L, Huttenlocher D, Kleinberg J, Lan X (2006) Group formation in large social networks: membership, growth, and evolution. In: Proceedings of the 12th ACM SIGKDD international conference on knowledge discovery and data mining (KDD '06). ACM, New York, NY, pp 44–54
5. Bishop CM (2006) Pattern recognition and machine learning (information science and statistics). Springer, New York, Secaucus, NJ
6. Cartwright D, Harary F (1956) Structural balance: a generalization of Heider's theory. Psychol Rev 63:277–293
7. Chen D, Fu Y, Shang M (2009) An efficient algorithm for overlapping community detection in complex networks. In: WRI global congress on intelligent systems, 2009 (GCIS '09), vol 1, pp 244–247
8. Chen Y, Wang X, Yuan B, Tang B (2014) Overlapping community detection in networks with positive and negative links. J Stat Mech Theory Exp 2014(3):P03021
9. Choudhury D, Paul A (2013) Community detection in social networks: an overview. Int J Res Eng Technol 2(2):83–88
10. Doreian P (2004) Evolution of human signed networks. Metodoloski Zvezki 1(2):277–293
11. Doreian P, Mrvar A (2009) Partitioning signed social networks. Soc Networks 31(1):1–11
12. DuBois T, Golbeck J, Srinivasan A (2011) Predicting trust and distrust in social networks. In: 2011 IEEE third international conference on privacy, security, risk and trust (PASSAT) and 2011 IEEE third international conference on social computing (SocialCom) (PASSAT/SocialCom 2011), Boston, MA, 9–11 October 2011, pp 418–424
13. Gomez S, Jensen P, Arenas A (2009) Analysis of community structure in networks of correlated data. Phys Rev E 80(1):16114
14. Guimerà R, Sales-Pardo M, Amaral LAN (2004) Modularity from fluctuations in random graphs and complex networks. Phys Rev E 70(2):25101
15. Hasan MA, Chaoji V, Salem S, Zaki M (2006) Link prediction using supervised learning. In: Proceedings of SDM 06 workshop on link analysis, counterterrorism and security
16. Javari A, Jalili M (2014) Cluster-based collaborative filtering for sign prediction in social networks with positive and negative links. ACM Trans Intell Syst Technol 5(2):24:1–24:19
17. Jin D, Yang B, Baquero C, Liu D, He D, Liu J (2013) Markov random walk under constraint for discovering overlapping communities in complex networks. Comput Res Repository abs/1303.5675
18. Kernighan BW, Lin S (1970) An efficient heuristic procedure for partitioning graphs. Bell Syst Tech J 49(2):291–307
19. Klamma R (2013) Community learning analytics – challenges and opportunities. In: Wang J-F, Lau R (eds) Advances in web-based learning – ICWL 2013. Lecture notes in computer science, vol 8167. Springer, Berlin, Heidelberg, pp 284–293
20. Kleinberg JM (1999) Authoritative sources in a hyperlinked environment. J ACM 46(5):604–632

21. Kunegis J, Lommatzsch A, Bauckhage C (2009) The slashdot zoo: mining a social network with negative edges. In: Proceedings of the 18th international conference on world wide web (WWW '09). ACM, New York, NY, pp 741–750
22. Kunegis J, Gröner G, Gottron T (2012) Online dating recommender systems: the split-complex number approach. In: Proceedings of the 4th ACM RecSys workshop on recommender systems and the social web (RSWeb '12). ACM, New York, NY, pp 37–44
23. Lancichinetti A, Fortunato S, Kertész J (2009) Detecting the overlapping and hierarchical community structure in complex networks. New J Phys 11(3):33015
24. Lerman K, Intagorn S, Kang J, Ghosh R (2011) Using proximity to predict activity in social networks. Comput Res Repository abs/1112.2755
25. Leskovec J, Huttenlocher D, Kleinberg J (2010) Predicting positive and negative links in online social networks. In: Proceedings of the 19th international conference on world wide web (WWW '10). ACM, New York, NY, pp 641–650
26. Li H, Zhang J, Liu Z, Chen L, Zhang X (2012) Identifying overlapping communities in social networks using multi-scale local information expansion. Eur Phys J B 85(6):190–198
27. Liben-Nowell D, Kleinberg J (2003) The link prediction problem for social networks. In: Proceedings of the twelfth international conference on information and knowledge management (CIKM '03). ACM, New York, NY, pp 556–559
28. Liu C, Liu J, Jiang Z (2014) A multiobjective evolutionary algorithm based on similarity for community detection from signed social networks. IEEE Trans Cybern 44(12):2274–2286
29. Mitchell TM (1997) Machine learning, 1st edn. McGraw-Hill, New York, NY
30. Newman MEJ (2003) Mixing patterns in networks. Phys Rev E 67(2):026126
31. Newman M (2004) Detecting community structure in networks. Eur Phys J B 38(2):321–330
32. Newman MEJ, Girvan M (2004) Finding and evaluating community structure in networks. Phys Rev 69:026113
33. Page L, Brin S, Motwani R, Winograd T (1999) The pagerank citation ranking: bringing order to the web. Technical report. Stanford InfoLab. http://ilpubs.stanford.edu:8090/422/
34. Palla, G., Derényi I, Farkas I, Vicsek T (2005) Uncovering the overlapping community structure of complex networks in nature and society. Nature 435(7043):814–818
35. Pothen A, Simon HD, Liou K-P (1990) Partitioning sparse matrices with eigenvectors of graphs. Eur Phys J B 11(3):430–452
36. Shahriari M, Jalili M (2014) Ranking nodes in signed social networks. Soc Netw Anal Min 4(1):1–12
37. Shahriari M, Klamma R (2015) Signed social networks: link prediction and overlapping community detection. In: The 2015 IEEE/ACM international conference on advances in social networks analysis and mining (ASONAM), Paris
38. Shahriari M, Krott S, Klamma R (2015) Disassortative degree mixing and information diffusion for overlapping community detection in social networks (DMID). In: Proceedings of the 24th international conference on world wide web companion, WWW 2015 - companion volume, pp 1369–1374
39. Shahriari M, Li Y, Klamma R (2016) Analysis of overlapping communities in signed complex networks. In: International conference on knowledge technologies and data-driven business (i-KNOW)
40. Shahriary SR, Shahriari M, Noor RM (2014) A community-based approach for link prediction in signed social networks. Sci Program J, 2015(5):10
41. Shen H-W (2013) Exploratory analysis of the structural regularities in networks. In: Community structure of complex networks, Springer theses. Springer, Berlin, Heidelberg, pp 93–117
42. Stanoev A, Smilkov D, Kocarev L (2011) Identifying communities by influence dynamics in social networks. Phys Rev E 84:046102
43. Šubelj L, Bajec M (2013) Model of complex networks based on citation dynamics. In: Proceedings of the 22nd international conference on world wide web companion WWW '13 Companion, pp 527–530

44. Symeonidis P, Iakovidou N, Mantas N, Manolopoulos Y (2013) From biological to social networks: link prediction based on multi-way spectral clustering. Data Knowl Eng 87:226–242
45. Tan P-N, Steinbach M, Kumar V (2005) Introduction to data mining, 1st edn. Addison-Wesley Longman, Boston, MA
46. Wang G, Shen Y, Ouyang M (2008) A vector partitioning approach to detecting community structure in complex networks. Comput Math Appl 55(12):2746–2752
47. Wu L, Wu X, Lu A, Li Y (2014) On spectral analysis of signed and dispute graphs. In: 2014 IEEE international conference on data mining (ICDM), pp 1049–1054
48. Xie J, Szymanski BK, Liu X (2011) Slpa: uncovering overlapping communities in social networks via a speaker-listener interaction dynamic process. In: 2011 IEEE international conference on data mining workshops (ICDMW), pp 344–349
49. Yang J, Leskovec J (2012) Community-affiliation graph model for overlapping network community detection. In: 2012 IEEE 12th international conference on data mining (ICDM), pp 1170–1175
50. Yang B, Cheung W, Liu J (2007) Community mining from signed social networks. IEEE Trans Knowl Data Eng 19(10):1333–1348

Chapter 4
Extracting Relations Between Symptoms by Age-Frame Based Link Prediction

Buket Kaya and Mustafa Poyraz

1 Introduction

Many complex systems, such as social, information, and biological systems, can be modeled as networks, where nodes correspond to individuals or agents and links represent the relations or interactions between two nodes. Network is a useful tool in analyzing a wide range of complex systems [1]. Many efforts have been made to understand the structure, evolution, and function of networks. Recently, the study of link prediction in network has attracted increasing attention [2]. Link prediction tries to infer the likelihood of existence of a link between two nodes, which has important theoretical and practical value. In theory, research on link prediction can help us understand the mechanism of evolution of complex network. In practical application, link prediction can be used in disease network to infer existence of a link so as to save experimental cost and time. It can also be utilized to online social networks to recommend friends for users, so as to improve the users' experience.

Link prediction problem has been interpreted and defined in many ways. All of these methods are based on the measures indicating the proximity between nodes. These measures proposed in the literature are generally categorized into semantic or topological/structural measures [3]. In semantic measures, the content of the nodes is taken into account to measure proximity. For example, in a co-authorship network, the similarity between keywords generated from published papers can be used to predict future connections among the authors. As apart from the semantic measures, the topological strategy employs the network structure to compute the proximity values. Topological measures are more widely adopted because they are more general and do not require detailed information related to the content of node. Moreover, this content is not always present depending on the social network

B. Kaya (✉) • M. Poyraz
Department of Electrical and Electronics Engineering, Fırat University, Elazığ, Turkey
e-mail: bkaya@firat.edu.tr; mpoyraz@firat.edu.tr

© Springer International Publishing AG 2017
J. Kawash et al. (eds.), *Prediction and Inference from Social Networks and Social Media*, Lecture Notes in Social Networks, DOI 10.1007/978-3-319-51049-1_4

considered. Topological measures are categorized into neighborhood-based or path-based measures [4]. The neighborhood-based measures consider the immediate neighbors of the nodes. According to this measure, two nodes are more likely to form a link if their sets of neighbors have a large overlap [3]. The most widely used measures among them are Common Neighbors [5], Preferential Attachment [5, 6], Adamic-Adar Index [7], and Jaccard's Coefficient [8]. The path-based measures generate a proximity value considering the paths between the related nodes. The main idea is that two nodes are more likely to form a link if there are short paths between the related nodes [9].

Many studies have been done to overcome the link prediction problem [2, 10, 11]. However, in order to predict new connections at future time, most of the previous studies are based on the applications of proximity measures to non-connected pairs of nodes in the network at present time. Proximity measures showing the similarity between pairs of nodes can be utilized either by unsupervised [2, 10, 12] or supervised link prediction [11, 13–15]. In unsupervised methods, a proximity measure is selected and utilized to rank node pairs in the network. The node pairs having good score are predicted to be linked. In the supervised approach, the link prediction problem is handled as a classification task and attributes of network are used as predictor attributes by a classification method [16–18]. A classifier employs these attributes to implement a binary classification to find whether the link will be occurred or not in the future. In all of the above-mentioned studies, the proximity values are calculated without considering the evolution of the network. In other words, the proximity measures are calculated using all network data up to the current network state without taking into account when links are created. Therefore, hidden and potentially valuable source of information is not adequately considered for link prediction.

In order to handle the above-mentioned problems, in our previous paper, we proposed a novel link prediction method with supervised strategy to identify the connections between diseases, building the evolving structure of the disease network with respect to patients' ages [17]. In this paper, we propose an unsupervised predictor to determine the risk of individuals to develop disease, and to undertake the correct actions at the earliest signs of illness. To this purpose, we first construct a weighted symptom network which indicates the relationships between abnormal parameters of disease. In this network, each node represents the abnormal parameters of patients; the edges connect these parameters appearing in the same patient. Next, we present a link prediction method based on the evolving cases of the constructed symptom network. Then, we increase the importance of more recent cases observed for a pair of symptom. Finally, we test the proposed method on the symptom network constructed with laboratory results of patients more than 210,000. The experimental results demonstrate the accuracy of our method on unsupervised prediction and encourage us for further analysis.

2   Evolving Symptom Networks

In this study, medical data comprise the results of the laboratory tests. In these data, there are many medical parameters (symptoms) tested for a patient. If all the values of these parameters are within reference interval, it can be said that the person coming to the hospital does not have a disease. However, if one of them or several parameters are not within reference interval, in other words, it exceeds the maximum value of the interval or it is lower than the minimum value of the interval, then this parameter exhibits an abnormal case and it is probable that the person having this laboratory result has a disease.

In the symptom network, nodes are the abnormal parameters which are not within reference interval. If the value of the parameter X is larger than the maximum value of reference interval, then this node is represented by $X.H$, where H stands for High. In contrast, if the value of the parameter X is lower than the minimum value of the reference interval, then it is represented by $X.L$, where L stands for Low. If the parameter X has a categorical or binary value, the unexpected value is represented by $X.A$ where A stands for Abnormal. In the constructed symptom network, edges connect these pairs of abnormal parameters appearing in the same patient. Thus, a relationship between two abnormal parameters exists whenever they appear simultaneously in a patient. Figure 4.1 shows a snippet of the symptom network. The patient medical records contain an important view regarding the co-occurrences of abnormal parameters affecting the same patient.

3   Proposed Method

As known, social networks are highly dynamic structures in which several connections and nodes tend to appear or disappear along time. The evolution of these networks brings valuable information about how connections tend to be formed,

Fig. 4.1 A snippet of the symptom network

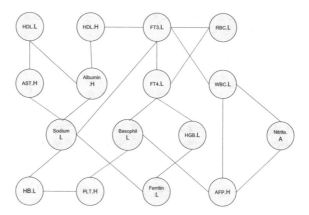

which will be then called as the link prediction task. In the present work, we propose a novel proximity measure that considers the evolving structure of a weighted network, such as symptom network, and evolving cases related to pairs of nodes in the network. The aim of the paper is to combine Homans' idea [19] that the strength of a connection between two nodes is directly associated with how often they interact with one another and Newman's method [20] that the bigger the number of common neighbors between two nodes, the higher is their probability to be connected in the future. Since the weighted network is used in this paper, we modified Newman's method in the form of "the larger the total weights of common neighbors between two nodes, the higher is their probability to be connected in the future."

3.1 The Evolving Structure of Symptom Network

The evolving structure represents states of the network at different age intervals in the past. For this purpose, in this paper, we split the network into several age-sliced snapshots. Then, we built the frames by grouping sequential snapshots. For instance, the frame at {[18–25]} age interval is a sub-graph containing all links observed among abnormal parameters of medical data of all people in between 18 and 25 years old. It should be noted that the size of each frame is the same as the length of the prediction window, predefined in the link prediction task.

3.2 The Evolving Cases

An evolving case is generally the action that leads a pair of nodes from a state (linked or unlinked) to another. An evolving case is defined as the increment or decrement of weight of a link between two nodes from a frame to its subsequent in a weighted network like symptom network. These cases can be categorized into one of the three different types: consistent, strengthening, and weakening.

3.2.1 Consistent Case

A consistent case in the weighted symptom network occurs when the weight of the relation between two nodes does not increase more than a case changing rate pre-determined or does not decrease less than this rate while the network evolves, that is when two abnormal parameters share a link in a frame and the strength of this connection is preserved in the subsequent one. In such a case, in order to consider a consistent case during the transition from the $(k-1)$-th to the k-th frame, $C_k(x, y)$ score for each pair of nodes (x, y) is defined as follows:

$$C_k(x, y) = \begin{cases} c \text{ if } w_{k-1}(x, y) - r.w_{k-1}(x, y) < w_k(x, y) < w_{k-1}(x, y) + r.w_{k-1}(x, y) \\ 0 \qquad\qquad\qquad\qquad\qquad otherwise \end{cases}$$

$$(4.1)$$

where $w_{k-1}(x, y)$ and $w_k(x, y)$ are the weights of connection between x and y nodes in frames F_{k-1} and F_k, respectively. r is the case changing rate pre-determined, the constant c ($0 < c \leq 1$) indicates the positive score for consistent cases since the strength of a tie between two nodes is preserved.

3.2.2 Strengthening Case

Strengthening cases occur when the weight of link between two nodes on different frames is increased more than the case changing rate pre-determined. In other words, they happen when the weight of two abnormal parameters in a frame increases more than the case changing rate in the next frame. The strengthening case $S_k(x, y)$ score associated with a pair (x, y) and a frame F_k is computed as follows:

$$S_k(x, y) = \begin{cases} s \text{ if } w_k(x, y) > w_{k-1}(x, y) + r.w_{k-1}(x, y) \\ 0 \qquad\qquad\qquad otherwise \end{cases}$$

$$(4.2)$$

where s ($0 < s \leq 1$) indicates a reward for the strengthening cases, its value should be positive since the tie between two nodes is strengthened.

3.2.3 Weakening Case

Weakening case is opposite to the strengthening one. In the kind of this case, the strength of existing link between two nodes decreases from a frame to its subsequent. The weakening case score $D_k(x, y)$ is computed as follows:

$$D_k(x, y) = \begin{cases} d \text{ if } w_k(x, y) < w_{k-1}(x, y) - r.w_{k-1}(x, y) \\ 0 \qquad\qquad\qquad otherwise \end{cases}$$

$$(4.3)$$

where d ($-1 \leq d < 0$) is a decrement value and should be negative since the strength of the connection between the nodes tends to decrease.

3.3 The Proximity Score in Evolving Symptom Networks

Many methods for link prediction compute scores to pairs of nodes by employing a chosen proximity measure. Thus, it is determined how similar those nodes are and, consequently, how likely a connection between them will be formed in a near future.

In this paper, we have combined the scores associated with primary cases, which are the evolving cases directly related to pairs of nodes under consideration, with the scores associated with secondary cases, which are the evolving cases observed in the nodes' neighborhood. The proximity score concerning a given pair of nodes (x, y) is determined as:

$$Score\,(x, y) = \sum_{k=2}^{n} [P\,(x, y, k) + \propto S\,(x, y, k)] \qquad (4.4)$$

$$P\,(x, y, k) = C_k\,(x, y) + S_k\,(x, y) + D_k\,(x, y) \qquad (4.5)$$

$$S\,(x, y, k) = \sum_{z \in \Gamma(x) \cap \Gamma(y)} P\,(x, z, k) + P\,(z, y, k) \qquad (4.6)$$

$P(x, y, k)$ computes the score of the case for the pair of nodes (x, y) observed in the transition from the frame $k-1$ to the frame k. The case observed at this moment may be one of three evolving cases: consistent, strengthening, or weakening. $S(x, y, k)$ gives the aggregated score of secondary cases concerning to the pair (x, y). Here, $\Gamma(x)$ is the set of neighbors of the node x in the constructed network. The parameter α is an amortization factor that shows how strong secondary cases affect the tie between x and y. In our experiment, the proposed measures were compared with different measures previously adopted in the literature for link prediction. We will use the proximity scores in the weighted networks since the edges are labelled with the number of patients having abnormal parameters in this paper. For this purpose, five proximity scores modified for weighted networks were considered: Common Neighbors (CN) [5], Jaccard's Coefficient (JC) [8], Preferential Attachment (PA) [6], Adamic-Adar Coefficient (AA) [7], and Resource Allocation Index (RA) [21].

3.4   The Algorithm

The steps of the proposed method are as follows:

1. The non-connected pairs or pairs having a weight less than pre-specified *threshold* are chosen from the validation set. However, the non-connected pairs that were directly or indirectly not affected by some case during the network evolution were eliminated.
2. The proximity scores of these pairs chosen are computed with respect to six different predictors (CN, JC, PA, AA, RA, and Our method).
3. The top 1000-pair having the highest proximity values are determined for testing. The top ranked pairs are considered as the future links that are more likely to appear.

4. If a non-connected pair in the validation set converts a connected pair in the
prediction frame or the weight of a pair having a weight less than predefined
threshold increases more than the case changing rate from the validation set to
the prediction frame, the pair of nodes is assumed as positive and assumed as
negative otherwise.
5. The performance measures of all the predictors considered are determined.

4   Experimental Results

We conducted a set of experiments using real medical data in order to evaluate
our proposed method. For this purpose, we first collected three different laboratory
results of the 210,134 patients. These are complete blood count, clinical biochemistry blood test, and urinalysis, which are the main sources for our study. Then,
we concentrate on predicting links between abnormal parameters of these tests for
various versions of the symptom network constructed. The proposed method was
implemented in Java.

In order to evaluate the performance of our method, the laboratory results
obtained from Fırat University Hospital in Turkey were used in our experiments.
Patients' medical records were taken suitably to the ethics rules of the hospital.
As mentioned earlier, while nodes represent the abnormal parameters, edges occur
when the couple of these abnormal parameters affect at least one patient. The total
number of nodes is 386 since if a parameter having continuous value is within
reference interval it is represented by two nodes as lower than or higher than normal.

The evolving structure of symptom network was initially built. In this construction, the frames at $[20, 25), [25, 30), [30, 35), [35, 40), [40, 45), [45, 50), [50, 55)$
age intervals were determined for the laboratory results and the snapshot at $[55, 60)$
age interval was used as the prediction frame. In this paper, we have employed
neighborhood-based measure because information about the connections around
nodes is important for us. The values c, s, d which are given for the consistent,
strengthening, and weakening cases, respectively, are estimated by evaluating the
link prediction performance on a validation set. The validation set is the last
frame before the prediction frame. Since the aim of the link prediction task is to
investigate a new link or weighted link in a network, the non-connected pairs or pairs
having a weight less than pre-specified *threshold* are chosen from the validation set.
However, the non-connected pairs that were directly or indirectly not affected by
some case during the network evolution were eliminated. Here, it should also be
noted that if a non-connected pair in the validation set converts a connected pair in
the prediction frame or the weight of a pair having a weight less than predefined
threshold increases more than the case changing rate from the validation set to the
prediction frame, the pair of nodes is assumed as positive and assumed as negative
otherwise. The *threshold* value is the proportion of weight of the node pair under
consideration to the sum of weights of all the links in the related frame. Throughout
this study, the *threshold* value is pre-specified as 0.5%. We also used the validation

Table 4.1 Precision values of six predictors in three different versions of the network

Network	CN	JC	PA	AA	RA	CBS
Unweighted	54.7	51.5	48.3	56.3	60.7	65.4
Weighted-1 (r = 20%)	**68.6**	**71.8**	**69.8**	**72.1**	**74.6**	**82.1**
Weighted-2 (r=40%)	62.5	68.5	65.1	67.2	67.3	73.6

Bold values show the best prediction values

Table 4.2 Recall values of six predictors in three different versions of the network

Network	CN	JC	PA	AA	RA	CBS
Unweighted	24.6	21.7	21.6	28.9	30.5	38.1
Weighted-1 (r = 20%)	**30.7**	**23.4**	**24.9**	**37.6**	**35.5**	**42.3**
Weighted-2 (r = 40%)	25.8	21.9	22.8	31.2	31.6	39.2

Bold values show the best prediction values

Table 4.3 F-Measures values of six predictors in three different versions of the network

Network	CN	JC	PA	AA	RA	CBS
Unweighted	33.9	30.5	29.8	38.1	40.5	48.1
Weighted-1 (r = 20%)	**42.4**	**35.2**	**36.7**	**49.4**	**48.1**	**55.8**
Weighted-2 (r = 40%)	36.5	33.1	33.7	42.6	43.0	51.1

Bold values show the best prediction values

set to empirically evaluate and select the most appropriate values of c, s, d, α, and r parameters in the link prediction task. We showed with some tests that the best performance was achieved at $c = 0.4$, $s = 0.9$, $d = -0.3$, and $\alpha = 0.07$. These results indicate that consistent (c) and weakening (d) values almost balance with each other. The best performance values were achieved when $c < s$.

The first experiment deals with finding the precision values of three different versions of the symptom network in order to compare our case-based scores to the traditional proximity scores modified for weighted networks. Table 4.1 reports the results of this experiment. As can be seen from Table 4.1, the best prediction performances are achieved by our link predictor (CBS) over weighted version 1 of the network. In this experiment, the unweighted version of the network obtains worse results. Similarly, the second experiment finds the recall values of every three versions of the network. Results in Table 4.2 show that while the weighted version 1 of the network exhibits worse performance at the traditional proximity measures, it obtains a better result at our method. The third experiment compares F-measure values for every three versions of the network. Our case-based scores (CBS) given in Table 4.3 obtained the best results from all the methods considered.

In the next experiment, the individual level risk of the symptom links are given as confidence values in Table 4.4. The first two columns in this table show the

Table 4.4 Confidence values of individual level risks at the top-5 relations found by our method

		Validation frame		Prediction frame	
Symptom$_1$	Symptom$_2$	1 → 2 (%)	2 → 1 (%)	1 → 2 (%)	2 → 1 (%)
AST.H	Cholesterol.H	1.3	1.1	5.9	5.8
Protein.H	Triglycerides.H	0.7	0.9	3.5	4.0
B12.L	Folic Acid.L	1.6	1.9	4.4	5.6
HGB.L	RBC.L	0.8	0.6	4.1	3.2
TSH.H	T3.L	0.6	1.4	2.5	6.6

frequency of the related symptoms in the validation frame. 1 → 2 (%), which is also called as confidence value of the rule *Symptom$_1$* → *Symptom$_2$*, represents the ratio of the number of patients having *Symptom$_1$* and *Symptom$_2$* (link weight) to the number of patients having *Symptom$_1$* in the validation frame. Similarly, 2 → 1 (%) is determined as:

$$\frac{\text{Link weight between Symptom}_1 \text{ and Symptom}_2}{\text{The number of patients with Symptom}_2}$$

For the first row, this means that while 1.3% of the patients having "High AST" at [50, 55) age interval also have "High Cholesterol," this rate raises to 5.9% at [55, 60) age interval. In other words, the risks of being captured both symptoms increases more than 4 times at [55, 60) age interval. This is an important finding for individual level of risk.

In the last experiment, we have raised the importance of more recent cases taking into account a monotonically increasing function in between the frames. For this purpose, we used a logarithmic function to weight recent cases more heavily than the old ones. Finally, the proximity score concerning a given pair of nodes (x, y) is modified as:

$$Score\,(x, y) = \sum_{k=2}^{n} I(k)\,[P\,(x, y, k) + \propto S\,(x, y, k)] \qquad (4.7)$$

I(k) represents the importance value of the frame. In this study, *I(k)* is *log(k)*.

Table 4.5 shows the precision values of the weighted version 1 of the network having three different importance in order to compare our case-based scores to the traditional proximity scores modified for weighted networks. As can be seen from Table 4.5, the best prediction performance is achieved by our link predictor with the importance of *log(k)*. In our experiments, we also observed a performance gain when the case-based score using $I(k) = log(k)$ was compared to the case-based scores using $I(k) = 1$ and $I(k) = k$ in the network. This situation reveals an important fact that recent cases carry more information about the occurrence of links.

Table 4.5 Precision values at different importance

Importance/Method	CN	JC	PA	AA	RA	CBS
I(k) = 1	68.6	71.8	69.8	72.1	74.6	82.1
I(k) = k	83.8					
I(k) = log(k)	86.2					

5 Conclusions

In this paper, we proposed a link prediction method for finding the most appropriate connections between abnormal parameters in symptom network. The method employs unsupervised machine learning strategy for link prediction. As the contributions of the paper, we first built a novel weighted symptom network which was combined from complete blood count, clinical biochemistry blood test, and urinalysis. Next, we propose an efficient link prediction method for weighted social networks, such as symptom network, by defining the evolving cases. In that network, the frames were constructed with respect to patients' ages.

The experiments conducted on real data sets illustrate that the proposed approach produces meaningful results and has reasonable efficiency. Our case-based link prediction score outperforms all the methods compared in this paper in terms of prediction, recall, and F-measure. The proposed method can also be directly applied to other diverse types of weighted social networks, or it can be extended to address problems not yet considered.

References

1. Kaya M, Alhajj R (2014) Development of multidimensional academic information networks with a novel data cube based modeling method. Inf Sci 265:211–224
2. Liben-Nowell D, Kleinberg J (2007) The link-prediction problem for social networks. J Am Soc Inf Sci Technol 58(7):1019–1031
3. Xiang EW (2008) A survey on link prediction models for social network data. PhD Thesis, The Hong Kong University of Science and Technology
4. Hasan MA, Zaki MJ (2011) A survey of link prediction in social network. In: Aggarwal CC (ed) Social network data analytics. Springer, pp 243–275
5. Newman MEJ (2001) Clustering and preferential attachment in growing networks. Phys Rev E 64:025102
6. Barabasi AL, Bonabeau E (2003) Scale free networks. Sci Am 288(5):60–29
7. Adamic LA, Adar E (2003) Friends and neighbors on the web. Soc Networks 25(3):211–230
8. Tan P-N, Steinbach M, Kumar V (2005) Introduction to data mining. Addison Wesley, Boston, MA, USA
9. Katz L (1953) A new status index derived from sociometric analysis. Psychometrika 18(1):39–43
10. Lu L, Zhou T (2011) Link prediction in complex networks: a survey. Phys A Stat Mech Appl 390(6):1150–1170

11. Hasan MA, Chaoji V, Salem S, Zaki M (2006) Link prediction using supervised learning, SDM'06: workshop on link analysis counterterrorism and security
12. Folino F, ve Pizzuti C (2012) Link prediction approaches for disease networks. ITBAM 2012:99–108
13. Lichtenwalter RN, Lussier JT, Chawla NV(2010) New perspectives and methods in link prediction. In: Proceedings of the 16th ACM SIGKDD international conference on knowledge discovery and data mining, pp 243–252
14. Sa HR, Prudencio RBC (2011) Supervised link prediction in weighted networks. In: Proceedings of the international joint conference on neural networks, pp 2281–2288
15. Soares PRS, Prudencio RBC (2013) Proximity measures for link prediction based on temporal events. Expert Syst Appl 40(16):6652–6660
16. Darcy D, Lichtenwalter RN, Chawla NV (2013) Supervised methods for multi-relational link prediction. Soc Network Anal Mining 3(2):127–141
17. Kaya B, Poyraz M (2015) Finding relations between diseases by age-series based supervised link prediction. In: Proceedings of the 2015 IEEE/ACM international conference on advances in social networks analysis and mining, pp 1097–1103
18. Gupta A, Sharma, S, Shivhare H (2016)Supervised link prediction using forecasting models on weighted online social networks, Volume 409 of the series Advances in Intelligent Systems and Computing, pp 249–261
19. Homans GC (1951) The human group. Routledge and Kegan, London
20. Newman MEJ (2001) The structure of scientific collaboration networks. In: Proceeding of the national academy of sciences of the United States of America, vol 98. pp 404–409
21. Ou Q, Jin Y.-D., Zhou T., Wang B.-H. and Yin B.-Q2007 Power-law strength-degree correlation from resource-allocation dynamics on weighted network. Phys Rev E752 (pt 1):021102. Lü, L., Zhou, T.: Link prediction in complex networks: a survey. Physia A: Statistical Mechanics and its Applications 390(6), 1150–1170 (2009).

Chapter 5
Link Prediction by Network Analysis

Salim Afra, Alper Aksaç, Tansel Õzyer, and Reda Alhajj

1 Introduction

In the recent surge and evolution of the world wide web, many opportunities arose for analyzing user-generated data, where the term big data became a buzz word and is now used almost everywhere such that high volume of data is available at all times in the Internet. Content of the available web-based data is mostly generated from on-line social networks and e-commerce web applications, among others. Domain specific data encapsulates either homogeneous or heterogeneous actors and the links connecting them leading to an n-mode network, where n is number of heterogeneous groups of actors. For instance, data generated from social networks relates mostly to interactions between users/visitors of the networks, where people are modeled as nodes and a friendship relationship is reflected as links connecting people. On the other hand, data generated from e-commerce websites models items (clothes, food, electronics, etc.) and people as nodes to reflect items viewed and bought by people. Accomplished purchase may suggest linking people to items. This behavior of interaction, whether between people or people and items, may be modeled as a social network. A social network can be viewed as a graph where a vertex represents a person or an item, and an edge corresponds to the underlying relationship between vertexes, e.g., friendship, collaboration, among others.

One of the attractive areas for network analysis is collaborations in research where researchers mostly coauthor papers reporting their findings. Collaboration between authors may last short or long leading to a number of coauthored papers over a period of time. Thus, collaboration may be modeled as a social network, and hence can be represented as a graph $G(V, E)$ where V is set of nodes or

S. Afra • A. Aksaç • T. Õzyer (✉) • R. Alhajj
Department of Computer Science, University of Calgary, Calgary, AB, Canada T2N 1N4
e-mail: salim.afra@ucalgary.ca; aaksa@ucalgary.ca; ozyer@etu.edu.tr; alhajj@ucalgary.ca

© Springer International Publishing AG 2017
J. Kawash et al. (eds.), *Prediction and Inference from Social Networks and Social Media*, Lecture Notes in Social Networks, DOI 10.1007/978-3-319-51049-1_5

vertexes representing authors and E is set of edges or links that exist only between researchers who have coauthored at least one paper. Building such a network or the similar, e.g., whether representing scholars or friends on Facebook or other networking sites, provides the possibility of analyzing and may be predicting or uncovering hidden links in the graph. The latter predictions may highlight a possible fruitful collaboration between potential researchers and hence would lead to a recommendation system (RS) which may bring to the attention of target researchers the importance of initiating a collaboration.

A recommendation system is an important mechanism which assists people in exploring items of their interest by guiding them into the specific set of available items in a system's directory. This kind of systems do their recommendations and predictions based on user's preferences and behavior. A separate profile is built for each user and items previously searched for or preferred by a user would help a recommender system in deciding what similar products to recommend. Recommendation systems are used in different domains and are very common in websites such as Google, Amazon, and other e-commerce websites in order to recommend to their users some suggested searches or guide them in buying new items. For example, Amazon recommendation system works by getting a list of items "user A" searched for and viewed, then uses this historical information, and checks what other users examined and purchased while also looking at the same set of items. After this step, the recommendation system will use the set of similar items for recommendations to the selected user. Recommendation systems are also used nowadays in social networking platforms such as Facebook and others to help in suggesting friends. This is done by predicting hidden links between actors and using some common features between users of social media. Such information may lead to new friendships between individuals in a social networking platforms such as Facebook and Twitter. In a similar settings, scientists are in need of different collaboration partners, i.e., experts in a specific topic similar to their research field. Indeed, research interests, co-citation, and bibliographic coupling have constituted some key metric and measure in searching for potential collaboration within a network.

Link prediction is also extensively used and important in the security domain. Since criminal activities occur in groups, finding a criminal may lead to identifying his/her whole criminal partners. Such that we can build a criminal network where nodes represent criminals and relationships represent an involvement of two criminals in an act. Performing link prediction in this kind of network will help governments, intelligence agencies, and other security companies to identify criminals and unveil possible actors involved.

Motivated by the above description, the work described in this paper tackles the issue of relating nodes in a general social network and then making appropriate suggestions by finding hidden links in the analyzed network. Completing this work will help in:

- Uncovering hidden relationships between nodes
- Categorizing and filtering the network
- Predicting links between nodes.

The method described in this paper has been tested on some benchmark networks. The reported results demonstrate the applicability and effectiveness of the proposed approach.

The rest of the paper is organized as follows: Section 2 reviews the most popular previous works. The proposed method is described in Sect. 3. Section 5 presents the conducted experiments, the evaluation process, and the results. Section 6 is conclusions and future research directions.

2 Related Work

A considerable amount of research work cover recommendation systems and link prediction, and how they may be applied in different fields. For example, in [1] the authors worked on rating prediction and recommendation of items for users. They carry out the ratings prediction by treating individual user-item ratings as predictors of missing ratings. The final rating is estimated by fusing predictions from the following sources: predictions based on ratings of same item by other users, predictions based on different item ratings made by same user, and ratings predicted based on data from similar users' ratings of similar items. Also in [2], the authors built an algorithm FolkRank ranking scheme which generates personalized rankings of items in a folksonomy and recommends users, tags, and resources. The basic idea is to extract communities of interest from folksonomy, which are represented by their top tags and most influential persons and resources. Once these communities are identified, interested users can find them and participate in. This way community members can more easily get to know each other by using link prediction. Furthermore, [3] proposed two new improved approaches for link prediction: (1) CNGF algorithm based on local information network and (2) KatzGF algorithm based on global information network.

There are also several efforts that investigate expert recommendation for business institutions, e.g., [2, 4]. Petry et al. [5] developed an expert recommendation system called ICARE, which recommends in an organization experts to work with. The focus of the work does not include relations between authors from their publications and citations, it rather considers organizational level of people, their availability, and their reputation. Reichling and Wulf [6] investigated effectiveness of a recommender system for a European industrial association in supporting their knowledge management, foregone a field study, and interviews with the employees. Experts were defined according to their collection of written documents which were automatically analyzed. Additionally, a post-integrated user profile with information about their background and job is used. Using bookmarking services of individual users in building user profiles provides further information about users' interests and confirms their recommendations.

Research on link prediction can also be found in [7], where authors proposed a supervised machine learning framework for co-offence prediction. The authors build a network of criminals and offenders first, then they started to find hidden links between known criminals and potential ones by relating offenders to socially related, geographically related, or experience related criminal cooperation opportunities. Additionally, Benchettara et al. [8] proposed a new link prediction algorithm to predict links in large-scale two-mode social networks. Based on topological attributes introduced in the paper, the score (or likelihood) of a link between two nodes can be measured, and they defined link prediction as a two class discrimination problem. Thus, a supervised machine learning approach is applied using these attributes to learn a prediction model. Finally, they validated their results on real datasets which are DBLP bibliographical database and bipartite transaction graph of an on-line music e-commerce site. Hasan et al. [9] developed another successful work using supervised learning for prediction; BIOBASE and DBLP networks are used to validate the model.

Another domain of link prediction in the research domain is to recommend possible future partnership to authors who never worked together before. Using this link prediction, the system will suggest people from other domains to work on similar projects, and this may lead to a fruitful partnership to the benefit of the community. However, the focus of the research is mostly focused on homogeneous networks of authors. Brandão et al. [10] modeled a social network of authors for recommending collaborations in academic networks. They presented two new metrics for their social network, namely institutional affiliation aspect and the geographic localization information. They analyzed how these metrics influence the resulting recommendations. Chen et al. [11] proposed a new way for scholar recommendations based on community detection. They used SCHOLST dataset in order to build a network of authors who are clustered into communities based on their research fields. Then they calculated friendship scores for each community in order to do coauthor recommendation based on communities. Davis et al. [12] introduced two approaches for link prediction in heterogeneous networks. In the first algorithm called unsupervised multi-relational link predictor (MRLP), they extended the well-known Adamic/Adar approach. Secondly, they used their previous research based on homogeneous networks in this study by extending for heterogeneous networks. A supervised framework for high performance link prediction (HPLP) shows that a supervised approach is superior to others, including MRLP. Tang et al. [13] proposed a methodology based on modularity analysis of heterogeneous YouTube dataset. Finally, Radivojac et al. [14] analyzed disease-gene networks.

Heck [15] focused on author link prediction, where authors are also modeled as nodes in a social network. What makes this work interesting is the selection of links between authors where bookmarking services are included in edge identification along with author co-citation and bibliographic coupling measurements. They argued how it is important to consider bookmarking along with the other metrics for better link prediction. Sun et al. [16] developed a methodology to predict coauthor relationship among authors in heterogeneous bibliographic network. They tested their algorithm on DBLP bibliographic network and according to their results prediction can be improved using logistic regression-based coauthorship prediction model based on meta path-based topological features. These are the combination of different meta paths and measures.

Discovering new hidden links in a social network is not a trivial task. In [17], when recommending new friendships in a traditional social network, the number of friends in common can be used to estimate the social proximity between users' ground model to smooth the rating predictions. Lichtnwalter and Chawla [18] showed how to evaluate developed methodologies in order to select the best technique. For more detailed information on this topic, the reader may refer to the review reported in [19, 20] (Table 5.1).

Table 5.1 The 11 similarity metrics used in link prediction; set $\Gamma(u)$ represents neighbors of node u in the network and $|\Gamma(u)|$ shows degree of node u

	Algorithm	Description
Adamic/Adar	$\sum_{z \in \Gamma(u) \cap \Gamma(v)} \frac{1}{log|\Gamma(z)|}$	This index measures similarity with counting of common neighbors z between nodes u and v by weighing the less-connected or rare neighbors more.
Jaccard	$\frac{|\Gamma(u) \cap \Gamma(v)|}{|\Gamma(u) \cup \Gamma(v)|}$	Common neighbors are divided by total number of neighbors of u and v. It looks for uniqueness in shared neighborhood.
Dice	$\frac{2|\Gamma(u) \cap \Gamma(v)|}{|\Gamma(u)| + |\Gamma(v)|}$	Common neighbors are divided by their arithmetic mean. It is a semimetric version of Jaccard.
Katz	$\sum_{\ell=1}^{\infty} \beta^\ell \cdot |paths_{u,v}^\ell|$	This index looks for path lengths and counts by weighting shorter paths between nodes more heavily. Parameter $\beta \in [0, 1]$ controls the contribution of paths and ℓ represents the length between nodes. Smaller values for β will decrease the contribution of higher values for ℓ.
Common neighbors	$|\Gamma(u) \cap \Gamma(v)|$	This index measures the number of shared neighbors.
Preferential attachment	$|\Gamma(u)| \cdot |\Gamma(v)|$	New connections are directly correlated with high degree of neighbors.
Salton	$\frac{|\Gamma(u) \cap \Gamma(v)|}{\sqrt{|\Gamma(u)| \cdot |\Gamma(v)|}}$	Common neighbors are divided by their geometric mean.
Resource allocation	$\sum_{z \in \Gamma(u) \cap \Gamma(v)} \frac{1}{|\Gamma(z)|}$	It is so similar to Adamic/Adar. While this index takes linear form, Adamic/Adar takes *log* form. But, this index is inversely more proportional to higher common neighbors.
Hub promoted	$\frac{|\Gamma(u) \cap \Gamma(v)|}{min\{|\Gamma(u)|, |\Gamma(v)|\}}$	Common neighbors are divided by minimum degree of neighborhood.
Hub depressed	$\frac{|\Gamma(u) \cap \Gamma(v)|}{max\{|\Gamma(u)|, |\Gamma(v)|\}}$	Common neighbors are divided by maximum degree of neighborhood.
Leicht–Holme–Newman	$\frac{|\Gamma(u) \cap \Gamma(v)|}{|\Gamma(u)| \cdot |\Gamma(v)|}$	Common neighbors are divided by square of their geometric mean.

3   The Methodology

3.1   The Algorithm

In our algorithm, we used centrality measures and path information to predict new links between nodes. Eigenvector centrality points popular nodes in a network. However, unpopular nodes (with not many connected links) may be more informative and discover strong links due to rarity in real networks. Betweenness centrality shows whether a certain part of a network is centralized or not. Centralized networks have a higher betweenness value since they have controller nodes to which everyone is connected. This situation will lead to less interaction between nodes because their connections will be over central nodes. Decentralized networks can have more shortest paths and can be more flexible. Also, we are not only looking for common neighbors while predicting new links to get information. The path-passed approach can provide more information compared to locally dealing with nodes in a network. Hence, close nodes will serve more possible connections, we are only considering shortest paths between nodes, meanwhile gaining from the complexity.

$$\sum_{z \in s.paths_{u,v}} \sum_{x \in V(z)} \frac{exp(-c_{eigen}(x)) \cdot c_{betw}(x)^{-1}}{length(z)} \qquad (5.1)$$

where nodes u and v satisfy $\{u, v \in V | e_{u,v} \in E\}$ in a given network $G(V, E)$. (V) and (E) are sets of vertices and edges, respectively. z is a shortest path, denoted $s.paths$, between u and v; $x \in V(z)$. $c_{eigen}(x)$ is eigenvector centrality of node x, and $c_{betw}(x)$ represents betweenness centrality of node x. $length(z)$ shows number of hops in path z between nodes u and v.

> **Case 1:** In Fig. 5.1a, shortest paths between 0 and 4 pass via nodes 2 and 6, while shortest paths from 0 to 3 pass via nodes 1 and 2. Node 2 is common for both cases. In this situation, the edge from 0 to 4 is more probable than the edge from 0 to 3 since node 6 has lower eigenvector and betweenness centralities compared to node 1. In our algorithm, we used $exp(x)$ function to avoid negative values and higher weight for rare neighbors. Also, the difference using exp is insignificant when the value is small, while it is larger when the value is large. The number of hops is 1 in this case since there is no shorter path of length 2.
>
> **Case 2:** In Fig. 5.1b, the edge from 0 to 2 is more probable than the edges from 0 to 3 and from 0 to 5 since it has many more shortest paths than the others.

3.2   Graph Database

In the implementation of our algorithm, we have used a graph database to store the graph network data. Traditionally storing data in a relational database dominated due to its high performance. But recently data types have changed in the Internet

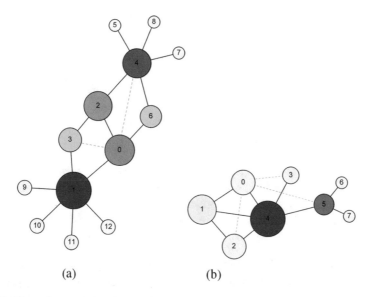

(a)   (b)

Fig. 5.1 Illustrative example of our prediction algorithm on a simple network. Size of nodes represents eigenvector centrality (larger size means higher value), while color of nodes shows betweenness centrality (*more reddish* means higher value). (**a**) Case 1. (**b**) Case 2

more towards social networking and big data domains; this involves complex interconnected information. Thus, storing and manipulating complex data has become an issue using traditional relational databases. This motivated for the development of several database structures like graph databases. A graph database provides a method or a tool to model and store a graph related data by focusing on the relationship between entities and attributes of the nodes as basic constructs of a data model [21].

We have used Neo4j tool which is an open source graph database based on Java combining graph storage services and a high performance scalable architecture for graph-based operations. In our work, we used Java libraries provided by Neo4j to create and store the datasets. We also used the graph-based methods in the library in order to get the shortest paths between nodes, eigenvalue, and betweenness centralities of each node.

4 Datasets

The best way to test our algorithm is to apply it on real-world networks to check if we can successfully predict links between real entities. Accordingly, we have applied our algorithm on six well-known real-world network datasets where the number of nodes and edges is shown in Table 5.2:

Table 5.2 Dataset networks
size

	# nodes	# edges
Zachary Karate club	34	78
Dolphin social network	62	159
Les Misérables	77	254
Books about us politics	105	441
Word adjacencies	112	425
American college football	115	613

- Zachary Karate Club [22]: is one of the most popular networks in terms of community structure. This network corresponds to members of a karate club at a US university in the 1970s; members are friends. This club has induced fights between its members such that members were split into half. This makes it a perfect real scenario network for link prediction on a member to check to which group he belongs.
- Dolphin Social Network [23]: is an undirected social network of frequent associations between 62 dolphins in a community living off Doubtful Sound.
- Les Misérables [24]: is a network corresponding for co-appearance of characters in the novel Les Misérables. It is interesting to test on this network as there are several communities to apply link prediction on them.
- Books About US Politics (orgnet.com): is a network of books about US politics sold by Amazon where edges between books represent frequent co-purchasing of books by same buyers.
- Word Adjacencies [25]: is an adjacency network of common adjectives and nouns in the novel David Copperfield by Charles Dickens.
- American College Football [26]: is a network of American football games between Division IA colleges during regular season in Fall 2000.

5   Experiments and Results

After collecting the datasets related to the various networks, the following steps are applied to run our algorithm which will output the confusion matrix for the evaluation code:

1. Randomly remove δ percentage of edges
2. Run the algorithms presented in Table 5.1 on the new network and get the corresponding confusion matrix
3. Calculate eigenvalue and betweenness centralities for all nodes in the network
4. Run our proposed algorithm
5. Select value α, which serves as a threshold for the algorithms predicted results.

The results shown in Tables 5.3, 5.4, 5.5, 5.6, 5.7, and 5.8 are average results where for each network we perform Step 1 of removing edges randomly ten times.

Table 5.3 Karate

	Accuracy			Sensitivity			Specificity			Precision			Miss rate			F1Score		
	I	II	III	I	II	III	I	II	III	I	II	III	I	II	III	I	II	III
Ours	0.99	0.99	0.97	0.11	0.10	0.18	0.99	0.99	0.98	0.04	0.11	0.16	0.89	0.90	0.82	0.06	0.09	0.11
Adamic	0.98	0.96	0.95	0.32	0.32	0.39	0.98	0.97	0.96	0.05	0.09	0.12	0.68	0.68	0.61	0.09	0.13	0.18
Jaccard	0.92	0.94	0.96	0.53	0.37	0.35	0.92	0.94	0.96	0.02	0.05	0.11	0.48	0.63	0.65	0.03	0.08	0.17
Dice	0.90	0.93	0.95	0.81	0.53	0.40	0.90	0.93	0.96	0.02	0.06	0.12	0.19	0.47	0.60	0.04	0.10	0.18
Katz1	0.85	0.87	0.89	0.77	0.69	0.66	0.85	0.88	0.89	0.01	0.04	0.08	0.23	0.31	0.34	0.03	0.08	0.14
Katz2	0.97	0.97	0.97	0.15	0.08	0.01	0.97	0.98	0.99	0.01	0.03	0.01	0.85	0.92	0.99	0.03	0.04	0.01
Katz3	0.97	0.97	0.97	0.00	0.00	0.00	0.98	0.98	0.99	0.00	0.00	0.00	1.00	1.00	1.00	0.00	0.00	0.00
Common neighbors	0.91	0.93	0.95	0.74	0.53	0.40	0.91	0.93	0.96	0.02	0.06	0.12	0.26	0.47	0.60	0.04	0.10	0.18
Preferential attachment	0.96	0.95	0.95	0.39	0.37	0.43	0.96	0.96	0.96	0.03	0.06	0.12	0.61	0.63	0.57	0.05	0.11	0.19
Salton index	0.90	0.93	0.95	0.81	0.53	0.40	0.90	0.93	0.96	0.02	0.06	0.12	0.19	0.47	0.60	0.04	0.10	0.18
Resource allocation	0.99	0.97	0.96	0.25	0.26	0.33	0.99	0.98	0.97	0.05	0.10	0.13	0.75	0.74	0.67	0.09	0.14	0.18
Hub promoted index	0.90	0.93	0.95	0.81	0.53	0.40	0.90	0.93	0.96	0.02	0.06	0.12	0.19	0.47	0.60	0.04	0.10	0.18
Hub depressed index	0.91	0.94	0.95	0.56	0.38	0.37	0.91	0.94	0.96	0.02	0.05	0.12	0.44	0.62	0.63	0.03	0.08	0.18
Leicht–Holme–Newman index	0.93	0.95	0.96	0.56	0.26	0.25	0.93	0.96	0.97	0.02	0.04	0.10	0.44	0.74	0.75	0.04	0.07	0.15

Table 5.4 AdjNoun

	Accuracy			Sensitivity			Specificity			Precision			Miss Rate			F1Score		
	I	II	III	I	II	III	I	II	III	I	II	III	I	II	III	I	II	III
Ours	1.00	1.00	0.99	0.05	0.09	0.09	1.00	1.00	1.00	0.01	0.04	0.05	0.95	0.91	0.91	0.01	0.06	0.06
Adamic	0.99	0.99	0.99	0.16	0.38	0.19	1.00	0.99	0.99	0.02	0.04	0.08	0.84	0.62	0.81	0.03	0.08	0.11
Jaccard	0.99	0.99	0.99	0.23	0.17	0.15	0.99	0.99	0.99	0.01	0.03	0.04	0.77	0.83	0.85	0.02	0.04	0.07
Dice	0.97	0.98	0.99	0.58	0.40	0.32	0.97	0.98	0.99	0.01	0.03	0.06	0.42	0.60	0.68	0.02	0.06	0.10
Katz1	0.98	0.98	0.98	0.60	0.66	0.54	0.98	0.98	0.98	0.01	0.04	0.06	0.40	0.34	0.46	0.02	0.08	0.10
Katz2	0.99	0.99	0.99	0.21	0.21	0.05	0.99	0.99	1.00	0.01	0.05	0.04	0.79	0.79	0.95	0.03	0.08	0.05
Katz3	1.00	1.00	1.00	0.00	0.00	0.00	1.00	1.00	1.00	0.00	0.00	0.00	1.00	1.00	1.00	0.00	0.00	0.00
Common neighbors	0.99	0.98	0.98	0.44	0.52	0.38	0.99	0.98	0.99	0.02	0.04	0.06	0.56	0.48	0.62	0.04	0.07	0.10
Preferential attachment	1.00	0.99	1.00	0.26	0.38	0.17	1.00	0.99	1.00	0.03	0.09	0.14	0.74	0.63	0.83	0.05	0.15	0.15
Salton index	0.97	0.98	0.99	0.70	0.41	0.36	0.97	0.98	0.99	0.01	0.03	0.06	0.30	0.59	0.64	0.02	0.06	0.10
Resource allocation	1.00	0.99	0.99	0.12	0.31	0.15	1.00	0.99	1.00	0.02	0.05	0.09	0.88	0.69	0.85	0.03	0.09	0.11
Hub promoted index	0.97	0.98	0.98	0.70	0.49	0.38	0.97	0.98	0.99	0.01	0.03	0.06	0.30	0.51	0.62	0.02	0.07	0.10
Hub depressed index	0.98	0.99	0.99	0.40	0.27	0.23	0.98	0.99	0.99	0.01	0.03	0.05	0.60	0.73	0.77	0.02	0.05	0.08
Leicht–Holme–Newman index	1.00	1.00	0.99	0.00	0.01	0.05	1.00	1.00	1.00	0.00	0.01	0.03	1.00	0.99	0.95	0.00	0.01	0.04

Table 5.5 Dolphins

	Accuracy			Sensitivity			Specificity			Precision			Miss Rate			F1Score		
	I	II	III	I	II	III	I	II	III	I	II	III	I	II	III	I	II	III
Ours	1.00	0.99	0.99	0.00	0.08	0.05	1.00	0.99	0.99	0.00	0.05	0.06	1.00	0.92	0.95	0.00	0.06	0.05
Adamic	0.96	0.97	0.98	0.63	0.58	0.31	0.96	0.97	0.98	0.02	0.07	0.10	0.38	0.42	0.69	0.04	0.13	0.15
Jaccard	0.97	0.97	0.98	0.50	0.50	0.29	0.97	0.98	0.99	0.02	0.08	0.11	0.50	0.50	0.71	0.04	0.13	0.16
Dice	0.96	0.97	0.98	0.63	0.58	0.31	0.96	0.97	0.98	0.02	0.07	0.10	0.38	0.42	0.69	0.04	0.13	0.15
Katz1	0.94	0.94	0.96	0.63	0.77	0.38	0.94	0.94	0.97	0.01	0.05	0.07	0.38	0.23	0.63	0.03	0.09	0.11
Katz2	0.98	0.99	0.99	0.13	0.10	0.03	0.99	0.99	0.99	0.01	0.04	0.02	0.88	0.90	0.98	0.02	0.06	0.02
Katz3	0.99	0.99	0.99	0.00	0.00	0.00	0.99	0.99	0.99	0.00	0.00	0.00	1.00	1.00	1.00	0.00	0.00	0.00
Common neighbors	0.96	0.97	0.98	0.63	0.58	0.31	0.96	0.97	0.98	0.02	0.07	0.10	0.38	0.42	0.69	0.04	0.13	0.15
Preferential attachment	0.91	0.93	0.96	0.75	0.69	0.34	0.91	0.93	0.96	0.01	0.04	0.05	0.25	0.31	0.66	0.02	0.07	0.09
Salton index	0.96	0.97	0.98	0.63	0.58	0.31	0.96	0.97	0.98	0.02	0.07	0.10	0.38	0.42	0.69	0.04	0.13	0.15
Resource allocation	0.97	0.97	0.98	0.56	0.48	0.31	0.97	0.97	0.98	0.03	0.07	0.10	0.44	0.52	0.69	0.05	0.12	0.15
Hub promoted index	0.96	0.97	0.98	0.63	0.58	0.31	0.96	0.97	0.98	0.02	0.07	0.10	0.38	0.42	0.69	0.04	0.13	0.15
Hub depressed index	0.96	0.97	0.98	0.56	0.58	0.31	0.96	0.97	0.98	0.02	0.08	0.10	0.44	0.42	0.69	0.04	0.14	0.15
Leicht–Holme–Newman index	0.99	0.99	0.99	0.06	0.23	0.13	0.99	0.99	0.99	0.01	0.10	0.11	0.94	0.77	0.88	0.02	0.14	0.12

Table 5.6 Football

	Accuracy			Sensitivity			Specificity			Precision			Miss Rate			F1Score		
	I	II	III	I	II	III	I	II	III	I	II	III	I	II	III	I	II	III
Ours	0.99	1.00	1.00	0.29	0.14	0.11	0.99	1.00	1.00	0.02	0.04	0.09	0.71	0.86	0.89	0.03	0.09	0.11
Adamic	0.99	0.99	0.99	0.82	0.75	0.57	0.99	0.99	0.99	0.02	0.07	0.13	0.18	0.25	0.43	0.04	0.13	0.18
Jaccard	0.99	0.99	0.99	0.72	0.74	0.57	0.99	0.99	0.99	0.04	0.09	0.14	0.28	0.26	0.43	0.08	0.08	0.17
Dice	0.99	0.99	0.99	0.82	0.75	0.57	0.99	0.99	0.99	0.02	0.07	0.13	0.18	0.25	0.43	0.04	0.10	0.18
Katz1	0.97	0.97	0.98	0.99	0.91	0.76	0.97	0.97	0.98	0.01	0.03	0.07	0.01	0.09	0.24	0.02	0.08	0.14
Katz2	1.00	1.00	1.00	0.67	0.41	0.11	1.00	1.00	1.00	0.04	0.11	0.08	0.33	0.59	0.89	0.08	0.04	0.01
Katz3	1.00	1.00	1.00	0.00	0.00	0.00	1.00	1.00	1.00	0.00	0.00	0.00	1.00	1.00	1.00	0.00	0.00	0.00
Common neighbors	0.99	0.99	0.99	0.82	0.75	0.57	0.99	0.99	0.99	0.02	0.07	0.13	0.18	0.25	0.43	0.04	0.10	0.18
Preferential attachment	0.96	0.97	0.97	1.00	0.99	0.92	0.96	0.97	0.97	0.01	0.03	0.04	0.00	0.01	0.08	0.02	0.11	0.19
Salton index	0.99	0.99	0.99	0.82	0.75	0.57	0.99	0.99	0.99	0.02	0.07	0.13	0.18	0.25	0.43	0.04	0.10	0.18
Resource allocation	0.99	0.99	0.99	0.81	0.73	0.57	0.99	0.99	0.99	0.02	0.07	0.13	0.19	0.27	0.43	0.04	0.14	0.18
Hub promoted index	0.99	0.99	0.99	0.82	0.75	0.57	0.99	0.99	0.99	0.02	0.07	0.13	0.18	0.25	0.43	0.04	0.10	0.18
Hub depressed index	0.99	0.99	0.99	0.82	0.75	0.57	0.99	0.99	0.99	0.02	0.07	0.13	0.18	0.25	0.43	0.04	0.08	0.18
Leicht–Holme–Newman index	0.99	0.99	0.99	0.82	0.74	0.53	0.99	0.99	1.00	0.02	0.08	0.18	0.18	0.26	0.47	0.05	0.07	0.15

Table 5.7 Les Misérables

	Accuracy			Sensitivity			Specificity			Precision			Miss Rate			F1Score		
	I	II	III	I	II	III	I	II	III	I	II	III	I	II	III	I	II	III
Ours	1.00	1.00	0.99	0.44	0.05	0.27	1.00	1.00	1.00	0.18	0.44	0.20	0.56	0.95	0.73	0.26	0.09	0.23
Adamic	0.99	0.99	0.99	0.80	0.61	0.60	0.99	0.99	0.99	0.05	0.13	0.20	0.20	0.39	0.40	0.09	0.21	0.30
Jaccard	0.98	0.99	0.99	0.76	0.64	0.47	0.98	0.99	0.99	0.03	0.10	0.15	0.24	0.36	0.53	0.06	0.17	0.23
Dice	0.97	0.98	0.98	0.80	0.72	0.58	0.97	0.98	0.99	0.02	0.07	0.14	0.20	0.28	0.42	0.04	0.13	0.23
Katz1	0.97	0.98	0.98	0.60	0.38	0.50	0.97	0.98	0.99	0.02	0.05	0.12	0.40	0.62	0.50	0.03	0.09	0.19
Katz2	0.99	0.99	0.99	0.60	0.39	0.24	0.99	0.99	0.99	0.04	0.10	0.16	0.40	0.61	0.76	0.08	0.16	0.19
Katz3	0.99	0.99	0.99	0.00	0.00	0.00	0.99	0.99	1.00	0.00	0.00	0.00	1.00	1.00	1.00	0.00	0.00	0.00
Common neighbors	0.99	0.99	0.98	0.76	0.55	0.61	0.99	0.99	0.99	0.05	0.13	0.14	0.24	0.45	0.39	0.09	0.21	0.23
Preferential attachment	0.98	0.99	0.98	0.52	0.34	0.49	0.98	0.99	0.98	0.02	0.06	0.11	0.48	0.66	0.51	0.05	0.10	0.18
Salton index	0.97	0.98	0.98	0.84	0.78	0.60	0.97	0.98	0.99	0.02	0.08	0.14	0.16	0.22	0.40	0.04	0.14	0.23
Resource allocation	0.99	0.99	0.99	0.72	0.58	0.54	0.99	0.99	0.99	0.06	0.15	0.22	0.28	0.42	0.46	0.11	0.24	0.32
Hub promoted index	0.97	0.98	0.98	0.88	0.79	0.60	0.97	0.98	0.99	0.02	0.08	0.14	0.12	0.21	0.40	0.04	0.14	0.23
Hub depressed index	0.98	0.98	0.98	0.76	0.64	0.56	0.98	0.98	0.99	0.03	0.08	0.14	0.24	0.36	0.44	0.06	0.15	0.23
Leicht–Holme–Newman index	0.99	0.99	0.99	0.20	0.28	0.20	0.99	1.00	0.99	0.03	0.12	0.13	0.80	0.72	0.80	0.05	0.17	0.16

Table 5.8 Polbooks

	Accuracy			Sensitivity			Specificity			Precision			Miss Rate			F1Score		
	I	II	III	I	II	III	I	II	III	I	II	III	I	II	III	I	II	III
Ours	1.00	1.00	1.00	0.14	0.05	0.07	1.00	1.00	1.00	0.09	0.19	0.11	0.86	0.95	0.93	0.11	0.08	0.09
Adamic	0.99	0.99	0.99	0.77	0.58	0.55	0.99	0.99	0.99	0.03	0.11	0.13	0.23	0.42	0.45	0.06	0.19	0.20
Jaccard	0.98	0.99	0.99	0.84	0.62	0.33	0.98	0.99	0.99	0.02	0.09	0.11	0.16	0.38	0.67	0.05	0.15	0.17
Dice	0.98	0.99	0.99	0.98	0.80	0.55	0.98	0.99	0.99	0.02	0.08	0.13	0.02	0.20	0.45	0.05	0.14	0.20
Katz1	0.98	0.98	0.99	0.77	0.72	0.62	0.98	0.98	0.99	0.02	0.06	0.10	0.23	0.28	0.38	0.04	0.11	0.17
Katz2	0.99	0.99	1.00	0.55	0.36	0.10	0.99	0.99	1.00	0.03	0.09	0.07	0.45	0.64	0.90	0.06	0.14	0.09
Katz3	1.00	1.00	1.00	0.00	0.00	0.00	1.00	1.00	1.00	0.00	0.00	0.00	1.00	1.00	1.00	0.00	0.00	0.00
Common neighbors	0.99	0.99	0.99	0.75	0.84	0.55	0.99	0.99	0.99	0.03	0.08	0.13	0.25	0.16	0.45	0.06	0.14	0.20
Preferential attachment	0.98	0.98	0.98	0.59	0.62	0.54	0.98	0.98	0.98	0.01	0.04	0.07	0.41	0.38	0.46	0.03	0.08	0.12
Salton index	0.98	0.99	0.99	0.98	0.83	0.55	0.98	0.99	0.99	0.02	0.08	0.13	0.02	0.17	0.45	0.05	0.14	0.20
Resource allocation	0.99	1.00	0.99	0.73	0.47	0.42	0.99	1.00	0.99	0.04	0.14	0.15	0.27	0.53	0.58	0.08	0.21	0.22
Hub promoted index	0.98	0.99	0.99	0.98	0.84	0.55	0.98	0.99	0.99	0.02	0.08	0.13	0.02	0.16	0.45	0.05	0.14	0.20
Hub depressed index	0.98	0.99	0.99	0.84	0.71	0.40	0.98	0.99	0.99	0.02	0.08	0.11	0.16	0.29	0.60	0.04	0.14	0.18
Leicht–Holme–Newman index	0.99	0.99	1.00	0.86	0.31	0.09	0.99	1.00	1.00	0.03	0.08	0.10	0.14	0.69	0.91	0.06	0.13	0.09

Fig. 5.2 Clustering
coefficient analysis of
datasets

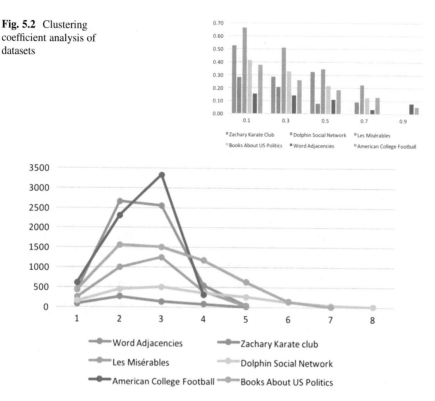

Fig. 5.3 Path distance distribution of datasets

Moreover, we chose three different values of δ for removing 10, 30, and 50% of the edges. We decided on these values based on clustering coefficient analysis. This analysis may help in finding missing links between nodes, called structural holes. High degree nodes will have lower local clustering values in this analysis which means more structural holes will exist in the network so that central nodes will collect all the flow of information and reduce alternative paths. In Fig. 5.2, we can see that we got more less local clustering values for 0.7 and 0.9 to the edge removal percentage. In addition, we experimentally used $\alpha = 0.1$ in all tests as threshold value for accepting an edge as predicted. We did this testing for the networks by choosing $\ell = 5$, which is the maximum depth used in all algorithms to search for a shortest paths between two nodes. In Fig. 5.3, we figured the path distance distribution of the datasets to show why we chose 5 as maximum depth. As shown for all networks, the performance of the algorithms are close to each other but our algorithm most of the times reports better values for the evaluation metrics than the other algorithms. In the various tables, *I* means $\delta = 0.1$, *II* means $\delta = 0.3$, and *III* means $\delta = 0.5$.

According to these tables, our precision and F1-score values are better than others in many cases by considerable margins. However, it is hard to decide which algorithm is better than others from accuracy results since the results are close to each other by small fraction. Also, we can clearly see from specificity results that we are not predicting non-existing links since our results are the higher compared to others, and our dataset is imbalanced which means that the number of negative examples ($FP + TN$, connections not to be predicted) is not close to the number of positive examples ($TP + FN$, connections to be predicted). Because of this, our sensitivity values are low since we have many false negatives by having more positive examples than negative examples.

To further check into how our algorithm is functioning and the advantage of using social network analysis in investigating the results of link prediction, we show in Fig. 5.4 a sample run of our algorithm on the Karate network with 30% of its edges removed. After removing edges, we ran the evaluation metrics we presented above to investigate the behavior of different algorithms compared to ours. In order to explain how our algorithm is predicting different than others, we used the color of the nodes to represent betweenness values on a white–red scale where white corresponds to low betweenness while red represents high betweenness. We also used size of the nodes to represent eigenvalues where the size of a node is directly proportional to its eigenvalue.

After running the algorithms on this network, we show edges which were successfully predicted by our algorithm as blue dashed lines; these edges connect nodes 32–33, 8–33, 15–32, and 20–32. All other algorithms have predicted the link between nodes 32 and 32 except Leicht–Holme–Newman algorithm. This is because these nodes have a large number of common neighbors (5) facilitating the

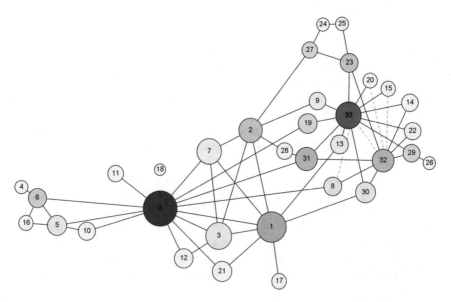

Fig. 5.4 Example of our prediction algorithm on the Karate Network with 30% of the edges removed

prediction of this edge. While none of the other algorithms predicted the existence of the other edges which were successfully predicted by our algorithm, except for Katz which predicted the edge between nodes 8 and 32. This reported result is due to the fact that there is no common neighbors between nodes 8–33, 15–32, and 20–32. Thus, the other algorithms failed to predict these links. While our algorithm successfully predicts the mentioned links because it does not only use common neighbors between two nodes but also considers the sum of all shortest paths between the two nodes.

6 Conclusions

In this paper, we tackled the problem of predicting the existence of links in a graph by using network analysis. Finding hidden relationships between actors in a network has various advantages in predicting different future partnership, collaboration, etc., that based on the actors properties can help and accomplish a new trend in the research domain. It also provides the ability to unveil already existing links between people. For example, detecting series of related criminals for security reasons. By performing link prediction using social network analysis, we are able to benefit from existing graph theory algorithms that provide good analytical solutions to the problem. In our paper, we used a combination of shortest path, betweenness, and eigenvalue centralities for the link prediction algorithm. We showed with examples how our algorithm can perform better on real-world datasets than other link prediction algorithms which mostly focus on common neighbors for prediction. We will continue to expand the algorithm by incorporating various other features from the network like closeness, connectedness, etc. We will also investigate the possibility of developing a classifier to help in the process.

References

1. Wang J, De Vries AP, Reinders MJT (2006) Unifying user-based and item-based collaborative filtering approaches by similarity fusion. In: Proceedings of the 29th annual international ACM SIGIR conference on research and development in information retrieval. ACM, New York, pp 501–508
2. Hotho A, Jäschke R, Schmitz C, Stumme G (2006) Information retrieval in folksonomies: search and ranking. Springer, Berlin
3. Dong L, Li Y, Yin H, Le H, Rui M (2013) The algorithm of link prediction on social network. Math Probl Eng 2013:7 pp.
4. Heck T, Peters I, Stock WG (2011) Testing collaborative filtering against co-citation analysis and bibliographic coupling for academic author recommendation. In: Proceedings of the 3rd ACM RecSys' 11 workshop on recommender systems and the social web, pp 16–23
5. Petry H, Tedesco P, Vieira V, Salgado AC (2008) Icare. a context-sensitive expert recommendation system. In: ECAI'08, pp 53–58

6. Reichling T, Wulf V (2009) Expert recommender systems in practice: evaluating semi-automatic profile generation. In: Proceedings of the SIGCHI conference on human factors in computing systems. ACM, New York, pp 59–68
7. Tayebi MA, Ester M, Glässer U, Brantingham PL (2014) Spatially embedded co-offence prediction using supervised learning. In: Proceedings of the 20th ACM SIGKDD international conference on knowledge discovery and data mining. ACM, New York, pp 1789–1798
8. Benchettara N, Kanawati R, Rouveirol C (2010) Supervised machine learning applied to link prediction in bipartite social networks. In: 2010 international conference on advances in social networks analysis and mining (ASONAM). IEEE, New York, pp 326–330
9. Hasan MA, Chaoji V, Salem S, Zaki M (2006) Link prediction using supervised learning. In: SDM'06: workshop on link analysis, counter-terrorism and security
10. Brandão MA, Moro MM, Lopes GR, Oliveira JPM (2013) Using link semantics to recommend collaborations in academic social networks. In: Proceedings of the 22nd international conference on World Wide Web companion. International World Wide Web conferences steering committee, pp 833–840
11. Chen J, Tang Y, Li J, Mao C, Xiao J (2014) Community-based scholar recommendation modeling in academic social network sites. In: Web information systems engineering–WISE 2013 workshops. Springer, Berlin, pp 325–334
12. Davis D, Lichtenwalter R, Chawla NV (2011) Multi-relational link prediction in heterogeneous information networks. In: 2011 international conference on advances in social networks analysis and mining (ASONAM). IEEE, New York, pp 281–288
13. Tang L, Wang X, Liu H (2009) Uncoverning groups via heterogeneous interaction analysis. In Ninth IEEE international conference on data mining, 2009. ICDM'09. IEEE, New York, pp 503–512
14. Radivojac P, Peng K, Clark WT, Peters BJ, Mohan A, Boyle SM, Mooney SD (2008) An integrated approach to inferring gene–disease associations in humans. Proteins Struct Funct Bioinf 72(3):1030–1037
15. Heck T (2013) Combining social information for academic networking. In: Proceedings of the 2013 conference on computer supported cooperative work. ACM, New York, pp 1387–1398
16. Sun Y, Barber R, Gupta M, Aggarwal CC, Han J (2011) Co-author relationship prediction in heterogeneous bibliographic networks. In: 2011 international conference on advances in social networks analysis and mining (ASONAM). IEEE, New York, pp 121–128
17. Lopes GR, Moro MM, Wives LK, De Oliveira JPM (2010) Collaboration recommendation on academic social networks. In: Advances in conceptual modeling–applications and challenges. Springer, Berlin, pp 190–199
18. Lichtnwalter R, Chawla NV (2012) Link prediction: fair and effective evaluation. In: Proceedings of the 2012 international conference on advances in social networks analysis and mining (ASONAM 2012). IEEE Computer Society, New York, pp 376–383
19. Liben-Nowell D, Kleinberg J (2007) The link-prediction problem for social networks. J Am Soc Inf Sci Technol 58(7):1019–1031
20. Lü L, Zhou T (2011) Link prediction in complex networks: a survey. Physica A 390(6):1150–1170
21. Silvescu A, Caragea D, Atramentov A (2002) Graph databases
22. Zachary WW (1977) An information flow model for conflict and fission in small groups. J Anthropol Res 33(4):452–473
23. Lusseau D, Schneider K, Boisseau OJ, Haase P, Slooten E, Dawson SM (2003) The bottlenose dolphin community of doubtful sound features a large proportion of long-lasting associations. Behav Ecol Sociobiol 54(4):396–405
24. Knuth DE (1993) The Stanford GraphBase: a platform for combinatorial computing, vol 37. Addison-Wesley, Reading, MA
25. Newman MEJ (2006) Finding community structure in networks using the eigenvectors of matrices. Phys Rev E 74(3):036104
26. Girvan M, Newman MEJ (2002) Community structure in social and biological networks. Proc Natl Acad Sci 99(12):7821–7826

Chapter 6
Structure-Based Features for Predicting the Quality of Articles in Wikipedia

Baptiste de La Robertie, Yoann Pitarch, and Olivier Teste

1 Introduction

Context Over the last few years, numerous of collaborative and crowdsourcing platforms are developing where the internaut is solicited to enrich knowledge bases. Wikis, Question answering (Q&A) websites are well-known examples where any user can create and edit content. Wikipedia is probably the most popular collaborative portal where any one can contribute to the editions of articles. Since its launch in 2001, the collaborative platform has grown to almost 35 millions of articles in more than 280 languages, with 2 millions of French articles and close to 5 millions of English ones.[1] The collaborative process involves more than 55 millions of individuals and generates 10 millions of edits each month, representing roughly 10 edits per second. This continuously increasing production of data leads to various scientific challenges among which the automatic quality assessment of the generated content.

While the main strength of Wikipedia is to allow anyone to contribute to its content, the website, as well as any other crowdsourcing platform, is not free from potential pitfalls of such an open collaborative editing process. Doubtful or even radically poor quality contents such as hoaxes, publicity, disinformation, or even acts of vandalism are likely to be published and might be available for consultation for several weeks before being identified and corrected. A popular example has involved the journalist J. Seigenthaler in the Kennedy assassination;

---

[1]http://en.wikipedia.org/wiki/Wikipedia:Statistics

B. de La Robertie (✉) • Y. Pitarch • O. Teste
Université de Toulouse, IRIT UMR5505, F-31071 Toulouse, France
e-mail: baptiste.delarobertie@irit.fr; yoann.pitarch@irit.fr; olivier.teste@irit.fr

© Springer International Publishing AG 2017   115
J. Kawash et al. (eds.), *Prediction and Inference from Social Networks and Social Media*, Lecture Notes in Social Networks, DOI 10.1007/978-3-319-51049-1_6

a fake content appeared in the Wikipedia biography page of the former in 2005 and was spread to other websites like Answers.com and Reference.com. The erroneous content remained online for more than 5 months [5].

In the scientific domain, different families of solutions tackle the problem of assessing the quality of articles. A first common and widely used idea exploited by most of the structural-based solutions is the *mutual reinforcement principle* hold between articles and contributors. The postulate says that authoritative users produce high quality contents and that conversely high quality contents are generated by authoritative users. A second interesting family of indicators are those based on *the temporal aspect* of the edit process offered by Wikipedia. The edits rate of high quality articles being quite different from others as well as the edits life cycles. A third promising track which receives too little attention concerns the structural properties of *the co-edit graph* of the authors. As motivated in a previous work [8], indicators in the collaborative graph built from the edit process of Wikipedia can help in discriminating the articles quality. In this study, we extend this previous work in order to show the interest of mining temporal collaborative graphs in the Wikipedia Quality Assessment Task. To the end, we propose a generic model and a HITS-like algorithm that combine the three introduced fundamental quality features.

Motivations Besides the different features considered by the state-of-the-art approaches [1, 3, 4, 15], we motivate the need for considering the co-edit graph in our quality model by answering the following questions:

Are the co-edit graphs of top quality articles denser than of poor quality articles? To answer this question, we built the co-edit graph for each of these two categories[2] according to the following principles: editors having authored some contents in at least one article of the desired quality represent the set of vertices. It exists an undirected edge between two editors if they have co-edited at least an article. Edges are weighted by the number of articles the pair of author has co-edited normalized by the number of documents in the category. The answer to this question as well as others basic statistics on the graphs are shown in Table 6.1. The graph associated with high quality articles is 12 times denser than the one associated with poor quality articles.

Table 6.1 Some statistics about the co-edit graphs per article quality level

Type of articles	Poor quality	High quality
Articles	18,823	245
Editors	36,973	9110
Collaborations	369,844	546,273
Density (10^{-5})	0.53	6.8
Avg. degree	15	26
Med. degree	5	10

---

[2]As stated in Sect. 5.1 the Wikipedia Editorial Team Assessment has manually labelled 30K articles. This enables the possibility to build some statistics on these categories.

Because comparing the density of very different size graphs might be legitimately discussed, the average and the median degree are also reported. Nodes associated with users collaborating in high quality articles have twice as many neighbors than those associated with poor articles. Based on these observations, we can suspect that high quality articles involve much more collaborations than poor quality articles. Thus, considering the collaborative process between the editor's looks to be a promising track to assess the quality of an article.

Are the co-editors used to work together on top quality articles? Besides the density of the co-edit graphs, we would like to analyze the nature of the relations between the users in order to determine in what extent authors collaborate together. Our intuition is that the higher the quality of an article is, the more frequent the interactions between co-editors are. The percentage of edges per weight per category is shown in Fig. 6.1. For readability purpose and because we aim at emphasizing on the impact of high weights, weights lower than 20 were omitted. Our intuition is obviously confirmed since high weights significantly appear more in the graph associated with high quality articles. Such an observation may indicate that good articles are authored by editors who should rely upon themselves since they are used to work together. Thus, considering the strength of the relation between co-editors looks to be a promising track to assess the quality of an article.

Contributions We propose to exploit the *mutual reinforcement principle* hold between articles and wikipedians as well as the co-edit graph of the collaborators to produce a set of structural features discriminating the quality of the Wikipedia articles. In summary, we propose the following contributions:

- We propose an unsupervised quality model based on a mutual reinforcement principle that integrates the user interactions and generalize the state-of-the-art approaches;
- We instantiate our quality model by using the co-edit graph of the contributors to produce a set of discriminative features;

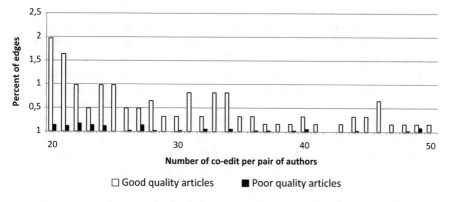

Fig. 6.1 Edges weight repartition for high quality articles (*white*) and poor quality articles (*black*)

- We incorporate this family of features in several standard classification and ranking models to show the interest of the structural-based features computed by the proposal;
- We evaluate our approach to show that it significantly performs better than the state-of-the-art approaches.

Paper Organization Related work is discussed in Sect. 2. Problematic is formally introduced in Sect. 3. Proposed quality model is presented in Sect. 4. Section 5 presents our experimental results.

2 Related Work

One of the earliest works [4] on Wikipedia has shown that the number of words in a document is a good predictor for article quality. More particularly, a specific class of articles in Wikipedia, known as *Features articles*, is easily predictable using the number of words in the article. However, this kind of model is limited to the articles written in English and is too specific for a particular class of articles. To overcome this limitation, De La Calzada et al. [7] propose to measure the quality of the articles using as many quality models as categories of articles. They first learn how to classify the articles using standard supervised learning (classification) techniques before applying the quality model associated with the predicted class. In practice, an SVM classifier is used to differentiate the "controversial" from the "stabilized" articles using a set of representative features (length of articles, number of internal links, etc.). Hence, a mixture of gaussian components (one per class) is used to compute the final quality of the articles. Other correlations between simple features such as the number of edits and unique editors are empirically demonstrated [15]. While these results are persuasive, both previous works consider the task as a binary classification problem where all non *Features articles* are considered as negative examples. To overcome these limitations, the lifespan of the edits has been widely used to quantify the quality of the articles. Adler et al. [1] show that the edits performed by the less authoritative users have a significantly higher probability of being quickly modified. This observation emphasizes the common idea that the age of an edit in Wikipedia is often considered as a good indicator of trustworthiness. Other works make use of the notion of *lifespan* to infer the quality or truthness of elements in the documents. For example, Adler et al. [1] introduce an approach that consists in measuring the reputation of the editors. The more an edit is preserved by subsequent reviewers, the more its author gains in reputation, and conversely, authors can lose reputation when their edits are reverted or deleted. The underlying intuition supposes that high quality contributions survive longer through the edit process due to an implicitly approvement of subsequent reviewers [3]. A study of Dalip et al. [6] shows that the structural features are, at least in Wikipedia, the most important family in a prediction task. More particularly, the non-consideration of this family of features leads to the greatest loss in terms of quality of the model.

Hu et al. [10] also model the authority of the authors and the reviewers to compute the quality of each word in the article. Three models based on the mutual dependency between articles quality and contributors authority are proposed such as the *Peer Review* model, which takes advantage of the implicit approbation of the reviewers, and the *Prob Review* model, which applies a decaying function over the authority of the reviewers around each approved word. Experiments show good results but are conducted over solely 200 documents. Moreover the structure between editors is not explicitly used. Adler et al. [2] also propose to credit each word with the reputation of the reviewers in proximity of the word and observe that words with low-trust assignment have high probability of being edited. In the work of Javanmardi et al. [11], the quality of an article is modeled as a time-dependent function, allowing quality of articles to evolve during time. In their work, only two states are considered making the study specific to the binary classification problem. Whöner et al. [16] propose to use the editing intensity during a period of time but experiments are conducted over solely 200 articles and consider only two classes : *Features articles*, i.e., good articles, and *articles for deletion*, i.e., low quality articles. A more recent work [14] reuses the mutual dependency concept between editors and texts and integrates the concept of lifespan as well as an adjustment of the authority of the reviewers in order to reduce the impact on text quality by vandal edits. A very recent work [12] explicitly formulates the mutual dependency between authors and articles with an Article-Editor network. The proposed model computes the quality of a document according to the editing relationships between article nodes and editor nodes using a variant of the PageRank algorithm [12]. Experiments are performed over only two classes of articles and none consider the relations between users. Finally, Suzuki [13] implicitly uses the structure of the co-edit network via the *h-index* measure in order to calculate the authority of the editors, but the quality of an article is directly computed using solely the derived authorities. To conclude, the author emits some doubts about the uniqueness of a quality function to distinguish good from bad articles.

Structural-based approaches of the literature are summed up in Table 6.2 according to the type of relations they consider. In the previous works, a mutual dependency principle and some surfaces and temporal features are used to discriminate high quality articles. But it seems that no one has explicitly combined these aspects and integrating co-edit relations between authors and reviewers. Moreover, most of these works consider the task of assessing the quality of Wikipedia article as a binary classification problem whereas six different grades have been proposed

Table 6.2 Related approaches

References	Considered relations		
	Authors/Articles	Reviewers/Articles	Authors/Reviewers
[1, 12, 14]	✓		
[2, 10, 18]	✓	✓	
[13]	✓	✓	✓

by the Editorial Team of Wikipedia. Our work can interestingly deal with a non-predetermined number of classes by producing a set of structural-based features that can be directly used in a ranking problem.

3   Problem Formulation

Let X be the feature space used to model the articles and $\mathbf{x}_i \in X$ be the d-dimensional feature vector associated with the article i. We denote by $\mathbf{x}_i^k$ the value of the k-th features. Let $y_i \in Y$ be the label of quality of the article i and $\hat{y}_i \in Y$ the predicted quality. We suppose that Y defines user preferences over the set of articles.[3] The objective is to find a function $f \in \mathcal{H}$, with $\mathcal{H}$ being the hypothesis space, such that the ranking induced by the predicted scores $f(\mathbf{x}_i) = \hat{y}_i, \forall i$ best respects the user preferences. In order to show the interest of considering the structural-based features produced by our quality model in a prediction task, three families of features are considered to characterize each article. More formally, the feature space X is separated into three subspaces:

1. Content feature space, denoted by X_C;
2. Temporal feature space, denoted by X_T;
3. Structural features space, denoted by X_S.

Therefore, each article can be mapped into the seven following configurations:

(A) $X = X_C$.
(B) $X = X_T$.
(C) $X = X_S$.
(D) $X = X_C \cup X_T$.
(E) $X = X_C \cup X_S$.
(F) $X = X_T \cup X_S$.
(G) $X = X_C \cup X_T \cup X_S$.

The instantiation is detailed in Sect. 5.2. It should be noted that the proposed quality model presented in the next section is aimed to feed the structural features space X_S.

4   Quality Model

The proposed quality model is grounded on the three following key features:

A circular relation, assuming that good articles are likely to be written by good editors, and conversely, good editors are more authoritative if they participate, by writing and reviewing good articles [2, 8, 14]. Besides, we postulate that the more

---

[3] User preferences over the Wikipedia articles is given in Sect. 5.

authoritative users participate to the elaboration of an article, the more the article is likely to be of good quality. The intrinsic dependency between quality and authority leads to an interdependent pair of equations, where the quality of an article is defined over the qualities of its individual pieces of contents, and the authority of a user over the individual pieces of content he/she authored and *approved*.

A word persistence assumption, postulating a gradual increase of the quality of a piece of content as new edits are performed [1, 16]. If many authoritative users who widely participate in an article leave previous edit on place, it is because they judge it as good quality. Thereby, the amount of new contributions of the reviewers should be catched in order to enhance the quality of long lasting pieces of content.

A collaborative work, assuming that good articles are likely to be the result of a grouping of expert users [8]. Our algorithm is designed to capture, from the co-edit graph of the collaborators, in what extent expert users are used to collaborate together. Reviews between these authors are very expressive because probably much more *trustworthy*. As observed in Fig. 6.1, the cumulative sum of edges weight of good quality articles is greater than those of poor quality articles. As it will be empirically demonstrated, it will be sufficient to push upward the scores of high quality articles.

4.1   Notations

Let A be the set of N articles and U the set of M users. Each article i is modeled by a set of *char sequences* $S_A^i = \{x_i^k\}_{1 \leq k \leq n_i}$, where n_i is the number of sequences in the article. Let $S = \cup_{1 \leq i \leq N} S_A^i$ be the set containing all the char sequences of the articles. The set of sequences a user j has authored is noted S_U^j. We note $\phi : S \rightarrow U$ the mapping function that returns the *distinct* author(s) of the char sequence(s) given as parameter. It should be noted that $\phi(S_U^j) = j, \forall j \in U$.

Example 1 Let A_{toy} be a set of 4 articles authored by U_{toy} a set of 4 editors. Corresponding char sequences S_{toy} are illustrated in Fig. 6.2. Author of each char

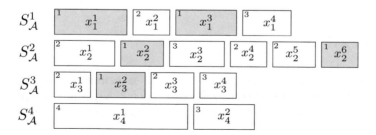

Fig. 6.2 Four articles and their respective char sequences. For instance, article S_A^4 (*bottom*) is composed of $n_4 = 2$ sequences authored by users $\phi(x_4^1) = 4$ and $\phi(x_4^2) = 3$, respectively

Fig. 6.3 Representation of
the co-edit graph
$G_{toy} = (V_{toy}, E_{toy})$. Users are
represented by *circles* (*left*)
and char sequences with
squares (*right*). Relation E is
represented in *black*. In *gray*
figures the relation which
associates to each user the
char sequences he/she
authored

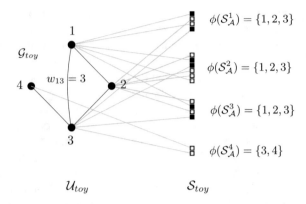

sequence is indicated on the top left corner of each sequence. For instance, the article
1 is composed by 4 sequences $S_A^1 = \{x_1^1, x_1^2, x_1^3, x_1^4\}$ authored by users 1, 2, 1, and
3, respectively, and we have $\phi(S_A^1) = \{1, 2, 3\}$. The first one x_1^1 (e.g., the beginning
of the article) has been authored by user 1, and the set of sequences that have been
authored by the user 1 (in gray in the picture) is $S_U^1 = \{x_1^1, x_1^3, x_2^2, x_2^6, x_3^2\}$.

Let $G = (V, E)$ be the undirect co-edit graph defined over the set of nodes $V = U$
and the relation $E \subseteq U \times U$. Weighted edges $w_{ij} \in \mathbb{R}^+$ between each pair of users
$(i, j) \in E$ measure the number of articles both users i and j have co-edited. Formally,
$w_{ij} = |\{k \in A : \{i, j\} \subseteq \phi(S_A^k)\}|$.

Example 2 Let $G_{toy} = (V_{toy}, E_{toy})$ be the co-edit graph constructed using the
previous example. A representation of G_{toy} is given in Fig. 6.3. The weight of the
edge between nodes 1 and 3 is $w_{13} = 3$ because associated users 1 and 3 have been
co-authors in three articles (articles 1, 2, and 3).

Finally, the *partial quality* score of a char sequence x_i^k is Q_i^k and the *partial authority*
score of the author of a sequence x_i^k is A_i^k. Global quality of an article j is noted Q_j.
The global authority of a user j is A_j. Score calculations are discussed in the next
sections.

4.2 Definitions

To capture the temporal interactions between the contributors, the notion of
lifespan of a sequence is introduced. It enables to formally define the concepts of
anterior and *posterior* neighborhoods of a char sequence and the notion of *implicit
approvement*.

Definition 1 (*Lifespan of a Sequence*) The *lifespan* l_i^k of a sequence x_i^k is the
number of revisions it survives until the latest version of the article.

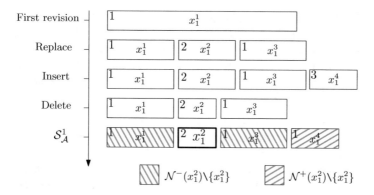

Fig. 6.4 Illustration of the edit history of article 1. Neighborhoods of sequence x_1^2 are computed considering the latest version of the article

Definition 2 (*Anterior Neighborhood*) The anterior neighborhood of a given sequence x_i^k, denoted by $N^-(x_i^k)$, is the set of sequences in the article i such that $\forall x_i^{k'} \in N^-(x_i^k)$, $l_i^{k'} \geq l_i^k$.

Definition 3 (*Posterior Neighborhood*) The posterior neighborhood of a given sequence x_i^k, denoted by $N^+(x_i^k)$, is the set of sequences in the article i such that $\forall x_i^{k'} \in N^+(x_i^k)$, $l_i^{k'} \leq l_i^k$.

Definition 4 (*Approved Sequence*) The sequence x_i^k is an approved sequence w.r.t. the user j (the user j *has approved* the sequence x_i^k) if the user j has authored at least one sequence in the posterior neighborhood of x_i^k, i.e., if $\exists x_i^{k'} \in S_U^j \cap N^+(x_i^k)$. We say that user j is a *reviewer* of sequence x_i^k.

Example 3 These four preliminary's concepts are illustrated inn Fig. 6.4. It gives an example of the edit process of the article 1 from the initial commit of the user 1 (top of the figure) to the latest revision (bottom of the figure). Authors of the sequences are indicated on the top left corner of each sequence. Let us consider the sequence x_1^2 in the latest version of the article. Since it appears three revisions before the latest version of the article (the sequence x_1^2 has been authored by the user 2 in the second revision), its lifespan is 3, i.e., $l_1^2 = 3$. The anterior neighborhood of the sequence x_1^2 is $N^-(x_1^2) = \{x_1^1, x_1^2, x_1^3\}$ and its posterior neighborhood is $N^+(x_1^2) = \{x_1^2, x_1^4\}$.

4.3 Model

Global Computation We propose to model the mutual reinforcement principle between the quality of an article and the authority of its author as follows:

$$Q_i = \underset{x_i^k \in S_A^i}{\mathcal{H}} (Q_i^k) \text{ and} A_j = \underset{x_i^k \in S_U^j}{\mathcal{H}} (A_i^k) \tag{6.1}$$

where the n-ary aggregation function $\mathcal{H}$ summarizes the partial quality and authority scores. The intuition is simple: we consider that the *global quality* of an article is function of the *partial quality* of each individual piece of content it contains. In the same way, the *global authority* of an author should be aggregated from all the pieces of contents he authored. In this work, we suppose a linear combination of the partial quality/authority scores.

Partial Computation When an authoritative user writes a content which is successively reviewed and approved by authoritative users, it is likely to be of good quality. This is even reinforced when many reviewers approved it and even more *when these reviewers have widely participated to the article*, e.g., the user 1 in Fig. 6.4. The more a reviewer throws himself/herself in an article, the more he/she is susceptible to see new edits and perform modifications if he/she judges the quality insufficient. This assumption becomes stronger when the number of authoritative reviewers increases. In other words, the amount of contributions of each reviewer seems to be fundamental to measure the concept of *approvement*. To formalize this intuition, the final quality Q_i^k of an individual sequence x_i^k is expressed by an *approvement function* which generically reflects the *temporal weighting schemes* (or relations) between authors and reviewers. Formally, the quality of the sequence x_i^k is defined as:

$$Q_i^k = \sum_{x_i^{k'} \in N^+(x_i^k)} \underset{\phi(x_i^{k'}) \to \phi(x_i^k)}{K} \left(A_i^k, A_i^{k'} \right) \tag{6.2}$$

where $\underset{j \to i}{K} : \mathbb{R}^+ \times \mathbb{R}^+ \to \mathbb{R}^+$ is the generic *approvement* function quantifying the implicit approvement by the user j of the sequence authored by the user i. Two forms of functions will be discussed in Sect. 4.4. In a symmetric way, the authority A_j^k is based on the sequences' quality he/she approved and is defined as:

$$A_j^k = \sum_{x_i^{k'} \in N^-(x_i^k)} \underset{j \to \phi(x_i^{k'})}{K} \left(Q_i^k, Q_i^{k'} \right) \tag{6.3}$$

Again, K must be instantiated in order to capture the approvement of a sequence by its reviewer.

4.4 Approvement Functions

The genericity of our quality model lies on the definitions on the approvement act of a sequence by a reviewer. What family of functions should be considered to give a fair credit to a particular sequence? Does the reviewer authority should influence the quality of the approved sequence? Is the quality of the char sequence a time-dependent function with the reviewer? To express the two intuitions presented in introduction, the two following approvement functions between an author i and a reviewer j are introduced and discussed:

$$K^1_{j \to i}(a, b) = a + b \qquad (6.4)$$

$$K^2_{\lambda\,j \to i}(a, b) = (ab)^{\lambda(1-w_{ij})} \qquad (6.5)$$

Instantiating our model with the approvement function K^1 leads to a slight but crucial variation of the *Peer Review* model proposed by Adler et al. In the original formulation [1], the quantity Q^k_i is computed as the sum over the authority of the author and the authorities of each *distinct* reviewer who approved x^k_i. More formally, if $\{z_1, z_2, \ldots, z_p\}$ is the set of p distinct users (including the author herself/himself) who approved the sequence x^k_i, then the computed quality of the sequence is indeed:

$$Q^k_i = A_{z_1} + A_{z_2} + \ldots + A_{z_p} \qquad (6.6)$$

One can demonstrate that this expression is a result of a particular instantiation of our model.

Lemma 1 *Let z_1 be the author of the sequence x^k_i and $\{z_2, \ldots, z_p\}$ be the set of $p - 1$ distinct reviewers who have approved x^k_i (including eventually the author himself/herself). Under our model, the use of the approvement function K^1 leads to a linear combination of the authority of the reviewers:*

$$Q^k_i = \sum_{i=1}^{p} \alpha_i A_{z_i} = \alpha_1 A_{z_1} + \alpha_2 A_{z_2} + \ldots + \alpha_p A_{z_p} \qquad (6.7)$$

where $\forall j > 1$, $\alpha_j = |\{x^{k'}_i \in N^+(x^k_i) \cap S^j_U\}|$, i.e., number of sequences of user j belonging to the posterior neighborhood $N^+(x^k_i)$, and $\alpha_1 = |N^+(x^k_i)| + |\{x^{k'}_i \in N^+(x^k_i) \cap S^1_U\}|$.

Proof Let $N^+(x^k_i) = \{x^{k_1}_i, x^{k_2}_i, \ldots, x^{k_s}_i\}$ be the set of s sequences authored by a set $\{z_1, z_2, \ldots, z_p\}$ of p distinct reviewers who approved sequence x^k_i. With K^1 to compute the partial quality Q^k_i, it holds:

$$Q^k_i = \sum_{x^{k'}_i \in N^+(x^k_i)} K_{\phi(x^{k'}_i) \to \phi(x^k_i)}(A^k_i, A^{k'}_i)$$

$$= \sum_{x^{k'}_i \in N^+(x^k_i)} A^k_i + A^{k'}_i$$

$$= (s \cdot A^k_i) + A^{k_1}_i + A^{k_2}_i + A^{k_3}_i + \ldots + A^{k_s}_i$$

By regrouping the identical reviewers,

$$Q_i^k = \underbrace{s \cdot A_i^k + A_i^{k_1}}_{\alpha_1 A_{z_1}} + \underbrace{A_i^{k_2} + A_i^{k_3}}_{\alpha_2 A_{z_2}} + \ldots + \underbrace{A_i^{k_s}}_{\alpha_p A_{z_p}} \tag{6.8}$$

we finally obtained :

$$Q_i^k = \alpha_1 A_{z_1} + \alpha_2 A_{z_2} + \ldots + \alpha_p A_{z_p} \tag{6.9}$$

□

Hence, the *Peer Review* model is a particular case of our model assuming all α_i equals to 1 (no weighting scheme is considered between the reviewers). The K^1 instantiation of our quality model (of K^1 model for short) enables the quality of a sequence to increase with the number of authoritative reviewers who have approved the sequence and also with the amount of contributions α they have authored in the article. The desired intuition is captured: if many authoritative users who have predominantly participated to the article have approved a given sequence, its quality will fairly increase.

The co-edit relations between users are captured by approvement function K_λ^2 (model K_λ^2 in short). When w_{ij} is close to 0 (none or very few relations between author j and reviewer i), the quality of a sequence increases only when both authors and reviewers are authoritative. Hence, unlike K^1, K^2 causes the approvement of an unauthoritative user to be almost unconsidered (final quantity is bounded by $\max(a, b)$ because of the normalization of the authority score). Strong relations notably enable new registered users to rapidly gain in authority, and even more if they collaborate with authoritative users. To control the strength of the co-edit weights and consequently the quality scores, a parameter $\lambda \in [0, 1]$ is introduced. Intuitively, when λ is close to 1, the function is elitist and tends to disfavor isolated users. Conversely, when λ is close to 0, the function is permissive and tends to encourage unauthoritative users. The effect of λ on the performances of the model K_λ^2 will be empirically shown in Sect. 5.4.

4.5   Calculation

The system formulated by the interdependent pair of equations (6.2) and (6.3) is solved by an iterative process that consists in alternatively computing the authorities A and the qualities Q until the scores remain stable. More details about the theoretical computation of the associated eigenvalues problem can be found in [9]. The following generic process is used:

1. Initialize randomly both authorities and qualities scores
2. Compute quality scores with Eq. (6.2)

3. Compute authority scores with Eq. (6.3)
4. Normalize scores.

The last three steps are repeated until a convergence state is reached. Let $Q^{(t)} \in \mathbb{R}^N$, resp., $A^{(t)} \in \mathbb{R}^M$, be the vector of quality scores, resp., authority scores, at the tth iteration of the algorithm. The convergence is reached when $d(Q^{(t)}, Q^{(t-1)}) + d(A^{(t)}, A^{(t-1)}) < \epsilon$, where d is a distance function and ϵ aimed to control the convergence. In the experimentations, the L_2 norm was used as a distance measure. For $\epsilon = 10^{-3}$, the convergence is rapidly reached (less than 10 iterations). Details about the computation are given by the following algorithms:

Algorithm 1 Quality(A, H)

1: **for all** $i \in A$ **do**
2: $\quad Q_i^{(t)} \leftarrow 0$
3: $\quad$ **for all** $x_i^k \in S_A^i$ **do**
4: $\quad\quad Q_i^{(t),k} \leftarrow \sum_{x_i^{k'} \in N^+(x_i^k)} H_{\phi(x_i^{k'}) \rightarrow \phi(x_i^k)} \left(A_i^{k,(t-1)}, A_i^{k',(t-1)} \right)$
5: $\quad\quad Q_i^{(t)} \leftarrow Q_i^{(t)} + Q_i^{(t),k}$
6: $\quad$ **end for**
7: **end for**

Algorithm 2 Authority(U, H)

1: **for all** $j \in U$ **do**
2: $\quad A_j^{(t)} \leftarrow 0$
3: $\quad$ **for all** $x_i^k \in S_U^j$ **do**
4: $\quad\quad A_j^{k,(t)} \leftarrow \sum_{x_i^{k'} \in N^-(x_i^k)} H_{j \rightarrow \phi(x_i^{k'})} \left(Q_i^{k,(t-1)}, Q_i^{k',(t-1)} \right)$
5: $\quad\quad A_j^{(t)} \leftarrow A_j^{(t)} + A_j^{k,(t)}$
6: $\quad$ **end for**
7: **end for**

With a basic implementation of the anterior and posterior neighborhood, the quality computation is performed in $\mathcal{O}(Nn^2)$, where $n = \max_i |S_i^A|$ and the authority computation is performed in $\mathcal{O}(Msn)$ where $s = \max_j |S_U^j|$.

5 Experiments

This section is dedicated to the presentation of our results. We first present the Wikipedia dataset used for the experimentations and detail the different features used for the prediction task and related models. Quantitative results are then presented using both unsupervised and supervised approaches. Finally, a qualitative interpretation of two representative co-edit graphs is provided to support our conclusions.

5.1 Wikipedia Dataset

We used a set of English articles from Wikipedia that have been reviewed by the
Editorial Team Assessment of the WikiProject. Therefore, each article is labelled,
according to the WikiProject quality grading scheme, using one of the following
flags $Y = \{S, C, B, GA, A, FA\}$. The user preferences defined over Y satisfies:

$$FA \succ A \succ GA \succ B \succ C \succ S$$

The label S stands for *Stub Articles* (very bad quality articles with no meaningful
content) while *FA* stands for *Featured Articles* (complete and professional articles).
This scale is used as a ground truth for evaluation. It should be specified that labels
were given according to the latest version of an article, which make sense, since we
have crawled and proposed to predict the quality of this latest version. We developed
a crawler in Java to parse grades, topics, and articles (from its first revision to the
latest version). The history of edits is stored in a relational database. The raw dataset
represents almost 130 Gb of text data. Statistics over these data are summarized in
Table 6.3.

A diff tool was developed in Python to extract the set of sequences that
survive until the latest revision. During this preprocessing step, the *lifespan* of
each sequence is updated. For each pair of consecutive revisions, the sequences are
propagated, split, and/or removed according to the possible sequence operations an
editor can perform (replace, insert, and delete characters). This preprocessing step is
applied over a stratified random collection of the raw data of nearly 23,000 articles
(a fix number of articles per category is randomly selected). In summary, more than
110,000 distinct users have produced around 2.8 million sequences. The co-edit
graph built over this dataset is composed by more than 111,000 nodes and 5 millions
of edges. Statistics about our dataset and resulted co-edit graphs are synthesized in
Table 6.4.

Table 6.3 Raw dataset
statistics

Grade	y_i	Articles	Revisions	Authors	Gb
FA	5	611	765,917	174,178	29
A	4	67	32,700	7492	2
GA	3	462	398,757	112,266	15
B	2	1459	1,283,225	330,010	44
C	1	3382	1,521,416	423,021	32
S	0	26,736	1,164,376	236,715	7

Table 6.4 Dataset

Grade	FA	A	GA	B	C	S
Articles	245	51	346	1012	1946	18,823
Authors per article (mean)	61	37	37	46	41	7
Lifespan per article (mean)	275	166	154	141	126	12
Sequences (mean)	1114	963	809	695	439	40
Sequences length (mean)	78	44	86	85	88	73
Nodes (10^3)	9.11	1.41	8.54	28.3	48.1	36.9
Edges (10^3)	546	62	406	1505	2279	369

Table 6.5 Feature spaces

Family	Dimension	Feature description
X_C	1	Number of characters
	2	Number of words
	3	Number of distinct words
	4	Number of char sequences
	5	Number of distinct editors
X_T	6	Number of revisions
	7	Mean of elapsed time between two edits
	8	Variance of elapsed time between two edits
	9	Mean of lifespan of sequences
	10	Variance of lifespan of sequences
X_S	11	Quality model instantiated with K^1
	12,13,…,22	Quality model using K_λ^2, for $\lambda \in \{0, 0.1, \ldots, 1\}$

5.2 Articles Features

Feature descriptions are given in Table 6.5. Each article $i \in A$ is modeled within the seven configurations of feature spaces introduced in Sect. 3. Moreover, an instance of the function $f \in \mathcal{H}$, we consider the four following families of models:

1. Identity model f_k^{ID}
2. Bayesian model f^{NB}
3. Binary decision tree f^{DT}
4. Random forest model f^{RF}

The first family, f_k^{ID}, constitutes a family of 22 unsupervised models. It returns for each article i the score $\hat{y}_i = f_k^{ID}(\mathbf{x}_i) = \mathbf{x}_i^k$, i.e., the value of the feature of the k-th dimension. In particular, the vector of scores f_{11}^{ID} is the one returned by the K^1 model (see Table 6.5). The last three models f^{NB}, f^{DT}, and f^{RF} are three propositions of supervised models where f is aimed to be learned over a subset of representative data.

5.3 *Evaluation*

Metrics Performances are evaluated using both standard ranking and classification evaluation metrics. To evaluate the ranking, the *Normalized Discount Cumulative Gain at k* (NDCG@k) was used [17]. It computes a normalized score based on the degree of relevance of each document and a decaying function of their rank. Formally, let σ be the permutation over the set of articles which consists to rank the articles by decreasing order of predicted quality $f(\mathbf{x}_i)$, i.e., $\sigma(i)$ is the position of the article i in the induced list. Hence, the DCG@k is computed as follows:

$$DCG(\sigma, k) = \sum_{i=1}^{k} \frac{2^{y_{\sigma(i)}} - 1}{\log(1 + i)} \qquad (6.10)$$

where y_j is the score label of article j. The more relevant documents in top position, the highest the DCG. To compare ranking, the metric is normalized over the value of the optimal ranking. Formally, the expression of the NDCG@k is given by:

$$NDCG(\sigma, k) = \frac{DCG(\sigma, k)}{DCG(\sigma^*, k)} \qquad (6.11)$$

where σ^* is the optimal ranking over the documents. In our case, the degree of relevance of a document is directly associated with its label: from 0 for documents belonging to class S (poor quality articles) to 5 for documents in class *FA* (very good articles). A perfect ranking consists to place all Features Articles on top, then all articles belonging to class A, and so on. A score equals to 1 indicating a perfect ranking over the first k documents. Once a permutation σ has been induced by the scores computed by a given model f, one can split the returned list of documents into 6 intervals according to the repartition of articles per grade in the ground truth. If we note I_Y the interval (or set of indices) in which the articles belonging to class Y should figure out in the optimal ranking, the Recall metric for the class Y can be expressed as follows:

$$R(\sigma, Y) = \frac{\{i : \sigma(i) \in I_Y \wedge y_i = Y\}}{|I_Y|} \qquad (6.12)$$

A recall of 1 indicating a perfect classification.

Competitors We compare our models to the following competitors proposed by Adler et al. [1] :

- *Basic* model. The effect of the reviewers is not taken into consideration.
- *Peer* model. Reviewer effects are considered as important as the author ones. Final authority of an author is the sum of the authority of the distinct reviewers.
- *Prob* model. Authority of the reviewer is slightly decreasing with the distance between author and reviewer words. In the following experiments, the best

decaying function $f(d) = \frac{1}{\max(0,d-\beta)+1}$ in [1] was used. In the formulation, d is the distance between the words of the author and the closest word of the reviewer and β is a user-parameter to control the maximum distance over which authorities of reviewers are not fully considered. In the experiments, parameter β was set to 1000 (distance in characters), corresponding to the best run among different values of the parameter.

Methodology   For the unsupervised models (proposed solution f_k^{ID} and competitors), we simply rank the articles by decreasing order of predicted quality $\hat{y}_i$ before applying the NDCG@k and the Recall metrics over the induced permutations. The values of k for the NDCG metric are directly derived from the repartition of the articles quality in the dataset, i.e., cumulative sum beyond the number of articles per class, saying that $k \in \{245, 296, 642, 1654, 3600, 22{,}423\}$). For the Recall metric, we report the performances for each of the 6 classes of quality, i.e., first class FA is evaluated using the first 245 documents, second class A is evaluated using the next 51, and so on.

Concerning the supervised models (proposed solutions f^{NB}, f^{DT}, and f^{RF}), several k-cross validations were performed. More precisely, for each $k \in \{2, 3, \ldots, 40\}$, a stratified random sampling over $\frac{1}{k}$ of the data is used to train a model, and the other $1 - \frac{1}{k}$ of the unseen data is used for the evaluation. By definition, for each k, the process is repeated k times such that each bin of size $\frac{1}{k}N$ is used for training exactly once. Because the number of articles in each class varies with k for each k-cross validation, it would not make sense to compute the NDCG and the Recall metrics at fixed positions. Hence, we recomputed for each k-cross validation the appropriate positions with respect to k to evaluate the models for each updated classes. For example, for a 3-cross validation, $\frac{2}{3}$ of the data is used for testing. Hence, we report the NDCG value for $k \in \{\frac{2}{3}245, \frac{2}{3}296, \frac{2}{3}642, \frac{2}{3}654, \frac{2}{3}3600, \frac{2}{3}22423\}$. Finally, in order to show the interest of considering the structural features, we have trained seven models according to the seven combinations of feature spaces presented in Sect. 3. For each of these models, the mean of the metrics over the k-cross validations is reported for each class.

5.4   Quantitative Experiments

5.4.1   Unsupervised Models

Results associated with the NDCG and the Recall are summarized in Tables 6.6 and 6.7, respectively. Best two results per column are highlighted using bold text. To make the notation less cluttered, we will omit, in this section, the *ID* upper script for the identity models. Thus, f_{11} corresponds to the K^1 quality model and models from f_{12} to f_{22} to the K_λ^2 model instantiated with $\lambda \in \{0, 0.1, \ldots, 1\}$.

B. de La Robertie et al.

Table 6.6 NDCG@k for unsupervised models

Model	FA	A	GA	B	C	S
Basic	7.7	8.01	11.06	46.5	51.84	69.97
Peer	38.67	41.24	51.94	66.13	82.24	84.46
Prob	28.4	30.66	40.55	57.68	72.06	79.73
f_1	31.65	35.63	56.94	71.65	80.85	82.69
f_2	32.04	35.8	57.62	71.79	81.1	82.86
f_3	32.25	34.05	53.5	70.3	80.15	82.3
f_4	31.65	35.63	56.94	71.65	80.85	82.69
f_5	29.61	30.75	41.02	60.22	76.88	80.03
f_6	35.5	37.05	46.03	69.44	83.31	83.31
f_7	0	0	0	0	0.02	52.96
f_8	0	0	0	0.05	0.75	54.88
f_9	42.72	43.83	53.19	67.5	82.12	84.45
f_{10}	37.34	38.78	51.23	64.28	81.26	82.94
f_{11}	**74.92**	**77.35**	79.16	**81.17**	**89.49**	**93.12**
f_{12}	39.7	42.2	54.66	69.75	82	84.31
f_{13}	41.38	44.29	56.12	71.75	82.88	85.19
f_{14}	43.78	44.65	57.55	72.86	83.81	85.71
f_{15}	46.24	47.42	60.61	75.05	85.24	87
f_{16}	50.16	51.25	65.1	77.91	86.45	88.27
f_{17}	64.55	66.5	77.28	80.61	**87.12**	91.21
f_{18}	70.19	73.27	**80.65**	**81.95**	86.07	92.7
f_{19}	66.74	67.12	72.48	74.39	84.76	90.58
f_{20}	11.29	11.31	13.37	28.4	66.23	72.43
f_{21}	10.86	11.5	20.09	52.67	66.7	74.63
f_{22}	**80.57**	**82.45**	**80.36**	77.42	83.63	**93.42**

Good Quality Articles Regarding to the classes FA, GA, and B, experiments are very conclusive with respect to the two metrics. Indeed, for each of these families of articles, it exists at least one proposed model that outperforms the competitors. In particular, f_{22} (or K_1^2) obtains the best results for the Features Articles for the two ranking and classification metrics. The co-edits weights of the reviewers seem to be interested indicators to discriminate very good articles. As expected, structural-based models K^1 and K_λ^2 outperform as well as the content-based and temporal-based models f_i for $i \leq 10$. Because we outperform the three independent solutions based on the number of edits f_6, the number of editors f_5, and the lifespan f_9 of the sequences, we demonstrate a real interest of combining these three features using K^1.

Poor Quality Articles Regarding to the class S and the Recall metric, best models are the state-of-the-art *Peer Review* model and the proposed one f_6 based on the number of revisions. The number of edits seems to be an important feature to discriminate bad quality articles. This makes sense since bad or incomplete articles are those that have not received so much attention by editors yet. The relatively bad discrimination of this class of articles by our models (recall around 94.5 for proposed model f_{22}, i.e., K_1^2) is a kind of cold start problematic. It might be explained by the presence, in the dataset, of very new articles that are written by very authoritative users: the content will automatically be considered as good quality (quality of a content directly derived from the authority of the author) and the proposed model will push these S articles upwards in the list whereas they should be considered as a poor quality articles (since they are not completed yet) and even if they are written by authoritative users.

Class A It is interesting to see that articles belonging to class A are very badly discriminated by every unsupervised models (see Table 6.7). The evolution of the mean rank and the dispersion of the articles belonging to class A for the different values of λ for the model K_λ^2 are plotted in Fig. 6.5 (right). Intuitively, the best model to discriminate the articles of class A would be the one the closest to the optimal solution in red: $K_{0.4}^2$ (f_{16}). But with respect to the Recall metric, the best discrimination of the articles of class A is obtained with $K_{0.3}^2$ (f_5); recall for class A is around 11%. For $\lambda \geq 0.5$, the co-edit relations seem to push backward the

Table 6.7 Recall per class for unsupervised models

Model	FA	A	GA	B	C	S
Basic	0.41	0	0.29	12.75	14.9	91.08
Peer	28.98	1.96	3.18	28.66	55.14	**98.94**
Prob	21.22	0	4.05	28.16	39.21	96.78
f_1	25.31	3.92	11.56	32.21	59.71	98.68
f_2	25.71	3.92	11.27	31.42	59.56	98.72
f_3	26.53	1.96	12.72	33	57.91	98.44
f_4	25.31	3.92	11.56	32.21	59.66	98.68
f_5	22.86	0	6.65	30.43	54.27	98.88
f_6	26.53	0	6.94	30.43	63.21	**98.92**
f_7	0	0	0	0	0.05	80.86
f_8	0	0	0	0.2	2.98	81.32
f_9	35.1	0	13.58	28.95	54.27	98.78
f_{10}	30.2	0	10.69	29.74	55.65	99.19
f_{11}	**67.35**	1.96	10.69	30.43	49.13	97.89
f_{12}	29.39	1.96	17.63	31.03	57.71	98.62
f_{13}	29.8	3.92	17.63	31.32	58.79	98.64
f_{14}	32.24	**11.76**	17.92	33.5	60.64	98.72
f_{15}	32.24	1.96	**18.79**	**35.18**	62.23	98.82
f_{16}	36.73	**5.88**	**22.83**	**37.55**	**62.33**	98.7
f_{17}	54.69	1.96	14.16	34.88	48.87	97.21
f_{18}	61.22	1.96	11.27	33.4	43.11	95.7
f_{19}	59.59	0	5.78	22.53	36.95	96.64
f_{20}	4.9	3.92	1.73	10.18	20.5	97.11
f_{21}	0.41	0	15.32	27.47	54.47	95.61
f_{22}	**75.51**	1.96	8.09	31.52	53.13	94.49

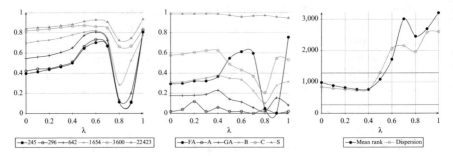

Fig. 6.5 At the *left*, evolution of the NDCG@k metric for the proposed model using approvement function K_λ^2 for λ increasing. At the *center*, evolution of the Recall metric for the proposed model using approvement function K_λ^2 for λ increasing. At *right*, evolution of the mean rank and the dispersion of the articles belonging to class A in the final ranking in function of λ for proposed model with approvement function K_λ^2. In *red*, the optimal mean rank (271). In *blue*, the mean rank returned by approvement function K^1 (1290)

articles of this class, making K^1 (in blue on the figure) more persuasive. Finally, still from Fig. 6.5, we see that it exists a value of λ which improves the precision and push upward this category of articles. This observation justifies the supervised approaches of the next section in order to automatically find this optimal value.

Parameter λ If the best two models for discriminating the FA articles are K^1 or K_1^2 in both Tables 6.6 and 6.7, best performances regarding the NDCG metric for middle classes are reached for middle values of λ. In particular, model K_λ^2 seems to discriminate mid quality articles more precisely for values of λ around 0.5. This behavior is illustrated in Fig. 6.5, which gives the evolution of the performances of K_λ^2 using the NDCG (left) and the Recall metric (center) in function of λ. An interesting collapse around 0.8 justifies the next supervised approaches. The evolution using the Recall metric is especially interesting since it indicates that the performances for the S class of articles are not affected so much by the λ parameter. Hence, in the case of an unsupervised scenario, considering $\lambda = 1$ seems to be beneficial to identify high quality articles while maintaining a satisfactory loss of quality for the other classes of articles.

5.4.2 Supervised Scenario

Results associated with the NDCG and the Recall metrics are summarized in Tables 6.8 and 6.9, respectively. For ease of reading, the three models f_{NB}, f_{DT} and f_{RF} are noted NB, DT, and RF, respectively. For the same reason, we denote by (a) the features space $X = X_C$, (b) the features space $X = X_T$ and so on, as noted in Sect. 3. Best two results per column are highlighted using bold text.

The interest of considering the proposed structural-based features is immediate: best results are achieved when the articles are represented with a combination

Table 6.8 Mean of the NDCG metric on the test sets for the different configurations of features for the three prediction models

X	f	FA	A	GA	B	C	S
(a)	NB	36.20	38.89	53.23	64.98	79.70	82.75
	DT	33.16	35.55	51.84	67.07	80.88	82.42
	RF	36.56	38.86	53.84	70.60	82.72	83.78
(b)	NB	23.39	24.20	30.89	42.93	58.59	74.25
	DT	36.54	38.61	51.17	65.54	82.94	82.94
	RF	45.65	47.47	59.19	72.94	86.43	86.45
(c)	NB	45.64	45.00	45.27	51.17	67.94	80.62
	DT	71.54	73.49	74.86	83.69	89.79	91.90
	RF	83.22	83.25	85.88	92.57	95.25	96.71
(d)	NB	41.12	43.95	54.22	67.17	82.29	83.97
	DT	40.22	42.58	55.79	69.33	84.32	84.33
	RF	50.08	51.96	64.36	77.18	87.99	88.01
(e)	NB	47.09	46.47	46.89	53.09	72.71	82.10
	DT	72.63	73.90	74.87	83.83	91.15	92.25
	RF	**84.58**	**84.22**	**86.28**	93.31	96.72	97.11
(f)	NB	46.90	46.43	47.00	51.91	70.05	81.32
	DT	72.63	74.17	75.22	84.65	92.49	92.49
	RF	84.51	84.10	86.27	**93.45**	**97.15**	**97.17**
(g)	NB	47.83	47.59	48.77	55.28	76.77	83.21
	DT	72.11	73.45	74.81	84.58	92.37	92.37
	RF	**84.86**	**84.35**	**86.40**	**93.86**	**97.24**	**97.26**

of features that include our structural-based features, i.e., cases (c) $X = X_S$, (e) $X = X_C \cup X_S$, (f) $X = X_T \cup X_S$, and (g) $X = X_C \cup X_T \cup X_S$. In particular, we observe that the Random Forest model achieves particular good results, with roughly 77% of precision for the class FA (see Table 6.9). It should be noticed that these performances represent the mean of the performances obtained for the 39 k-cross validations with k varying from 2 to 40. Hence, results reported in Tables 6.8 and 6.9 are quite pessimistic. For the best case ($k = 2$), half of the data (roughly 10,000 articles) is used for training and the other half for testing. But for the worst case, only $\frac{1}{40}$ of the data is used for training, which leads to a very unbalanced difficult scenario for small classes such as A and FA classes. Indeed, in this latter case, only two or three articles of class A are used during the training phase.

Details of the results for each k-cross validation for the FA class are plotted in Figs. 6.6 and 6.7. It corresponds to the evolution of the NDCG and Recall metric for the FA class of articles in function of the size of the train dataset. More precisely, we see the evolution of the performances of the three models in function of the number of folds k used during the k-cross validations. Interestingly, the Naive Bayes model is not sensitive to k for the three cases (a), (b), and (d), and the best performances for $k = 2$ are achieved using a combination of content and temporal features (NDCG is around 0.4 and Recall around 0.36). However, when we use a combination of the three feature spaces, the performances of the Naive Bayes model decrease while

Table 6.9 Mean of the Recall metric on the test sets for the different configurations of features for the three prediction models

X	f	FA	A	GA	B	C	S
(a)	NB	28.35	1.83	13.51	26.18	56.31	98.53
	DT	24.97	2.12	13.57	28.76	60.46	99.15
	RF	28.12	2.44	14.80	30.73	63.73	99.40
(b)	NB	12.77	2.18	12.87	26.91	46.20	97.49
	DT	29.00	2.05	13.61	28.88	61.79	99.99
	RF	36.15	2.49	14.76	30.49	64.43	99.99
(c)	NB	35.85	2.21	7.79	19.89	32.22	94.07
	DT	66.66	2.48	24.16	53.97	74.51	98.46
	RF	75.23	**9.53**	**45.63**	**68.84**	83.14	99.02
(d)	NB	33.59	1.98	15.39	28.53	59.41	99.17
	DT	32.20	2.22	14.65	29.61	64.20	**100.00**
	RF	40.18	3.53	16.51	32.66	68.17	99.98
(e)	NB	37.28	2.13	8.41	22.04	41.76	95.93
	DT	67.87	2.37	23.65	54.62	80.02	99.26
	RF	**76.91**	**9.06**	42.02	68.28	87.77	99.67
(f)	NB	37.15	2.18	9.58	21.05	41.22	95.46
	DT	67.88	2.48	23.85	57.36	86.27	**100.00**
	RF	76.61	8.94	**42.52**	**68.69**	90.38	99.98
	NB	38.34	2.24	9.34	23.90	47.25	97.32
(g)	DT	67.17	2.66	23.24	57.55	86.43	**100.00**
	RF	**77.18**	8.92	41.26	68.76	**90.85**	99.99

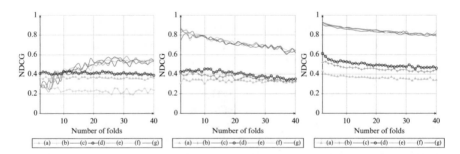

Fig. 6.6 Evolution of the NDCG metric for the dynamic FA class of articles on the test sets for the seven combinations of feature spaces for the three models NB (on the *left*), DT (in the *middle*), and RF (on the *right*) in function of the number of folds k used for training during the k-cross validation

the number of example for the training phase is increasing. Concerning the two other models, all measured performances improve in test with the number of training examples. The charts clearly indicate that the cases (c), (e), (f), and (g) are far away above the cases (a), (b), and (d). In particular, from $k = 40$ to $k = 2$, roughly 0.2 points are gained for the two DT and RF models for the two metrics. Once again, the use of the structural features clearly improves the performances of the models.

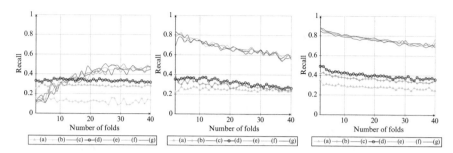

Fig. 6.7 Evolution of the Recall metric for the dynamic FA class of articles on the test sets for the seven combinations of feature spaces for the three models NB (on the *left*), DT (in the *middle*), and RF (on the *right*) in function of the number of folds k used for training during the k-cross validation

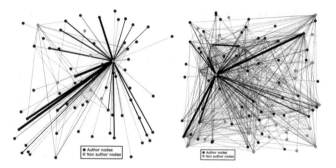

Fig. 6.8 The filtered co-edit graphs of the articles related to *A. Hillgruber* and *Kaga*

5.5   *Qualitative Interpretation*

We now present the two co-edit graphs associated with a representative poor quality article (see Fig. 6.8 (left)) and a top quality article (see Fig. 6.8 (right)) according to our proposed metric. The former is dedicated to *A. Hillgruber*,[4] a conservative German historian, and has been labelled as a C-class article by the Editorial Team Assessment because of its partisanship. The latter is dedicated to *Kaga*,[5] a Japanese aircraft career, and has been labelled as an FA-class. Both graphs have been obtained using the following methodology: vertices are the union of the authors and their respective co-editors (in all the dataset). For ease of reading and because very weak co-edit relations are not of interest in this study, edges with weight equals 1 have been filtered out as well as subsequent isolated vertices. Table 6.10 sums up some

---

[4]https://en.wikipedia.org/wiki/Andreas_Hillgruber
[5]https://en.wikipedia.org/wiki/Japanese_aircraft_carrier_Kaga

Table 6.10 Some statistics about the co-edit graphs of the articles related to *A. Hillgruber* and *Kaga*

	A. Hillgruber	Kaga
Vertices	42572	14023
Edges	128322	55368
Authors	63	58
Vertices (filtered)	70 (0.17%)	93 (0.66%)
Edges (filtered)	87 (0.07%)	420 (0.76%)
Class repartition (%)	FA: 4.3 / A: 0 / GA: 5.6 B: 1.4 / C: 85 / S: 3.7	FA: 55.7 / A: 0.8 / GA: 3.6 B: 25.3 / C: 8 / S: 6.6

statistics on these two graphs. In the following we will refer to the non-filtered graph related to *A. Hillgruber*, resp. *Kaga*, as $\mathcal{G}_C$, resp. $\mathcal{G}_{FA}$ and to its filtered version as $\mathcal{G}'_C$, resp. $\mathcal{G}'_{FA}$.

Since less than 1% of the edges has weights greater than 1, authors most often collaborate one time only whatever the quality of the article. This observation can though be sharpened by carefully analyzing the two graphs. On the one hand, $\mathcal{G}'_C$ shows the following characteristics: (1) it is very sparse, (2) very few edges connect two author nodes, and (3) very few non-author nodes remain in $\mathcal{G}'_C$ after the filtering step. On the other hand, $\mathcal{G}'_{FA}$ shows some opposite characteristics: (1) it is denser than $\mathcal{G}'_C$, (2) much more edges connect two author nodes, and (3) more non-author nodes remain in $\mathcal{G}'_{FA}$ after the filtering step. These observations confirm the soundness of our approach. Indeed, in poor quality article, the collaboration is punctual only; the added value of reviews is thus lower for top quality articles where authors are more used to collaborate. Additionally, we compute the class of articles in which authors mostly participate. The repartition per class is given in the last line of Table 6.10. This shows that top quality articles, resp. low quality articles, are mostly written by authors who are used to write such good quality articles, resp. low quality articles. This is a clear evidence of the pertinence of the proposed solution to integrate the structural properties of the co-edit graph.

Results confirm that combining different families of features and notably structural-based features from the Wikipedia co-edit graph is beneficial and makes the solution closer to the optimal solution. The co-edit graph is clearly discriminating to capture authoritative users and thus, good articles.

6 Conclusion

Crowdsourcing platforms provide the possibility for anyone to freely contribute to their publicly available content. One inherent drawback of this collaborative process is the emergence of poor quality content. In this paper, we tackled the problem of automatically assessing articles quality in the particular case of Wikipedia. We proposed a structural-based algorithm based on a mutual reinforcement principle that produces a set of efficient features that characterize the quality of the articles on Wikipedia. Our formulation generalized previous works by introducing the notion of *approvement functions*. Moreover, these introduced concepts integrate the co-edit graph and take advantage of the relations between the editors. Such a formulation facilitates the theoretical comparison with the state-of-the art approaches which can naturally be expressed as instances of our model. Motivated by some strong hints that legitimate the importance of considering the co-edit graph, two novel approvement functions were designed. The first function reinforces the quality of a content as a function of both the authority of the reviewers and the amount of their contributions in the article. The second function aims at capturing the relation between the authors and the reviewers since we have considered that the reviews of editors who are used to work together are more trustworthy. For this purpose, the co-edit network between editors was constructed and has appeared to have very interesting features. These features combined with a set of content and temporal-based features have shown to be of a real interest in a prediction task. In particular, experiments conducted both in a unsupervised and supervised scenario on real Wikipedia articles are very conclusive. The proposed model, by improving the state-of-the-art methods, empirically confirmed our two intuitions and opened several perspectives.

In future work, we first plan to extend our model by generalizing the notion of neighborhood. Indeed, we think it would be beneficial to consider both horizontal (time) and vertical (documents) aspects of the neighborhood of a sequence. Notably, such an operator would enable to even more generalize our model and to reformulate other state-of-the-art works, e.g., the *Prob Review* model. Scalability is a major concern. We plan to study the possibility to adapt our model to a Big Data environment using some parallelization strategies. Finally, because of the intensive edit rate of Wikipedia articles, adapting our model to a streaming environment would enable the quality calculation on the fly.

References

1. Adler BT, de Alfaro L (2007) A content-driven reputation system for the wikipedia. In: Proceedings of the 16th international conference on world wide web (WWW '07). ACM, New York, NY, pp 261–270
2. Adler BT, Chatterjee K, de Alfaro L, Faella M, Pye I, Raman V (2008) Assigning trust to wikipedia content. In: Proceedings of the 4th international symposium on wikis (WikiSym '08). ACM, New York, NY, pp 26:1–26:12
3. Biancani S (2014) Measuring the quality of edits to wikipedia. In: Proceedings of the international symposium on open collaboration (OpenSym '14). ACM, New York, NY, pp 33:1–33:3
4. Blumenstock JE (2008) Size matters: word count as a measure of quality on wikipedia. In: Proceedings of the 17th international conference on world wide web (WWW '08). ACM, New York, NY, pp 1095–1096
5. Cox LP (2011) Truth in crowdsourcing. IEEE Secur Priv 9(5):74–76
6. Dalip DH, Gonçalves MA, Cristo M, Calado P (2011) Automatic assessment of document quality in web collaborative digital libraries. J Data Inf Qual 2(3):14:1–14:30

7. De la Calzada G, Dekhtyar A (2010) On measuring the quality of wikipedia articles. In: Proceedings of the 4th workshop on information credibility (WICOW '10). ACM, New York, NY, pp 11–18
8. de La Robertie B, Pitarch Y, Teste O (2015) Measuring article quality in wikipedia using the collaboration network. In: Proceedings of the 2015 IEEE/ACM international conference on advances in social networks analysis and mining 2015 (ASONAM '15). ACM, New York, NY, pp 464–471
9. Golub GH, Van Loan CF (2012) Matrix computations, vol 3. JHU Press, Baltimore
10. Hu M, Lim E-P, Sun A, Lauw HW, Vuong B-Q (2007) Measuring article quality in wikipedia: Models and evaluation. In: Proceedings of the sixteenth ACM conference on conference on information and knowledge management (CIKM '07). ACM, New York, NY, pp 243–252
11. Javanmardi S, Lopes C (2010) Statistical measure of quality in wikipedia. In: Proceedings of the first workshop on social media analytics (SOMA '10). ACM, New York, NY, pp 132–138
12. Li X, Tang J, Wang T, Luo Z, de Rijke M (2015) Automatically assessing wikipedia article quality by exploiting article-editor networks. In: ECIR 2015: 37th European conference on information retrieval. Springer, Berlin
13. Suzuki Y (2015) Quality assessment of wikipedia articles using <i>h</i>-index. J Inf Process 23(1):22–30
14. Suzuki Y, Yoshikawa M (2013) Assessing quality score of wikipedia article using mutual evaluation of editors and texts. In: Proceedings of the 22Nd ACM international conference on conference on information & knowledge management (CIKM '13). ACM, New York, NY, pp 1727–1732
15. Wilkinson DM, Huberman BA (2007) Cooperation and quality in wikipedia. In: Proceedings of the 2007 international symposium on wikis (WikiSym '07). ACM, New York, NY, pp 157–164
16. Wöhner T, Peters R (2009) Assessing the quality of wikipedia articles with lifecycle based metrics. In: Proceedings of the 5th international symposium on wikis and open collaboration (WikiSym '09). ACM, New York, NY, pp 16:1–16:10
17. Yining W, Liwei W, Yuanzhi L, Di H, Wei C, Tie-Yan L (2013) A theoretical analysis of NDCG ranking measures. In: Proceedings of the 26th annual conference on learning theory
18. Zeng H, Alhossaini M, Fikes R, McGuinness L (2006) Mining revision history to assess trustworthiness of article fragments. In: Proceedings of the 2nd international conference on collaborative computing: networking, applications and worksharing

Chapter 7
Predicting Collective Action from Micro-Blog Data

Christos Charitonidis, Awais Rashid, and Paul J. Taylor

1 Introduction

The tendency of people to form groups and connect with others in order to communicate and engage in collective activities is associated with the human nature [14]. Such groups tend to expand and evolve over time gaining large number of members who share the same purpose or ideas. Social networks are considered the main means of individual recruitment and social movement's growth [12]. In the age of Internet, and particularly of social media, the collective construction of social movements has become much faster and effortless to achieve [8, 16].

The use of social media as a means of communication and organisation of collective action, and particularly as a means of mobilisation—in terms of "getting people to the streets"—has become a modern phenomenon. Recent and very relevant examples include London riots and Arab Spring. Social media played a key role, in the former, in arranging times and locations of rioting and looting, while in the latter, in most cases, in organising peaceful protests.

Detecting the signals of such unfolding phenomena has thus become a key topic of research in the last few years. Recently, several systems have been developed to cope with the problem of event detection in social media. For example, Twitcident [1] analyses information on Twitter for detecting incidents or crises. Similarly, TEDAS [20], a Twitter-based event detection and analysis system, detects new

C. Charitonidis (✉) • A. Rashid
Security Lancaster Research Centre, Infolab21, Lancaster University, Lancaster LA1 4WA, UK
e-mail: c.charitonidis@lancaster.ac.uk; a.rashid@lancaster.ac.uk

P.J. Taylor
Department of Psychology, Centre for Research and Evidence on Security Threats (CREST), Lancaster University, Lancaster LA1 4YF, UK
e-mail: p.j.taylor@lancaster.ac.uk

© Springer International Publishing AG 2017
J. Kawash et al. (eds.), *Prediction and Inference from Social Networks and Social Media*, Lecture Notes in Social Networks, DOI 10.1007/978-3-319-51049-1_7

events and analyses their spatial and temporal patterns. TwitterMonitor [22] automatically detects and analyses emerging trends on Twitter by identifying keywords that suddenly appear in tweets at an unusually high rate. Even though a range of approaches have been proposed, most of them focus on analysing the major themes or trending topics at a particular point in time (i.e., the strong signals). Consequently, their analysis is undertaken on the "sufficiently visible" topics, leaving only a few, or no possibilities, for the early detection of emerging events. Others rely on simple counting of keywords/messages, which can provide insufficient insights, or even miss key indicators of such dynamically evolving phenomena.

As yet, little attention has been paid to the identification, detection and analysis of the so-called weak signals. Weak signals are early indicators of emerging trends that initially appear at the fringes of mainstream discussion, but quickly and unexpectedly reach a tipping point. Such weak signals have been likened to "hardly discernible cracks anticipating an earthquake" [3]. Weak signals have been proposed as a means of early indicators [3], but no concrete study exists that identifies or validates that such weak signals exist and can be identified reliably.

In this paper, we focus on analysing Twitter data from the London riots in 2011 through a multi-disciplinary approach, drawing upon human behaviour and computer science, to identify and detect weak signals that could be early indicators of the subsequent widespread offline actions. The proposed approach can be of benefit to social scientists and crime analysts in studying online phenomena and their links to offline behaviours.

We used in our previous work [10] a range of techniques, including frequency, keyword, geo-spatial, semantic and sentiment analysis, to identify and detect patterns of weak signals in social media. This paper builds on the previous work, and goes beyond by training a machine learning model for predicting violent collective actions. A comprehensive feature extraction and selection process is performed to select the most informative features (predictors) for use in model construction. Following on from that, a series of experiments and comparisons between different machine learning algorithms is conducted to select the best performing model for predictions.

The rest of this paper is structured as follows. Section 2 reviews the most related work. Section 3 presents the proposed model for the detection of weak signals in social media. Section 4 gives a brief overview of the 2011 London riots and details the methods used for collecting and analysing the data. Section 5 presents the results of the data analysis. Section 6 provides a detailed description and presents experimental results of training and evaluating a machine learning model for tweets classification, based on the insights gained from the previous analysis. Section 7 gives a discussion of the results and highlights the weak signals identified in our analysis. Finally, Section 8 concludes the paper and provides directions for future work.

2 Related Work

Most relevant research on mining and analysis in online social media has focused on Social Network Analysis (SNA). SNA is a common approach used to study the relationships and information exchanged between actors (i.e., individuals or groups). It can reveal groups of actors within a network after examining the relationships existing between them [19]. Identifying how online groups are formed and developed, it is possible to gain predictive insights into future group behaviour. The patterns of relationships can reveal the way that two or more actors are linked to each other based on their in-between interactions, such as the quantity and frequency of information exchanged. Actors who exchange information frequently or exchange a large amount of information have stronger relationship than actors whose communication patterns are infrequent or low in the amount of information exchanged [19]. Tie strength allows analysts to examine key groups/actors in the network, as well as to assess the possibility that information will flow from one actor to another. For example, StakeNet [21] uses social network measures to identify and prioritise stakeholders in the network, i.e., individuals or groups that can influence or be influenced. Although strong ties tend to be more influential [26], weak ties should not be considered negligible as they may act as bridges filling a structural hole, i.e., the absence of link between two individuals or groups in the network [17]. These isolated actors can maintain relationships in more than one group and can transfer new and valuable information to other groups. Both strong and weak ties can play a key role in information diffusion, and therefore, their analysis can reveal possible indicators of dynamic changes in the group.

Bodendorf et al. [6] propose an approach for detecting opinion leaders and trends by combining text mining for the extraction of opinions and communication relationships among users, and social network analysis (using measures such as degree centrality, closeness centrality and betweenness centrality) for the identification of opinion leaders. Opinion leaders play a significant role when it comes to spreading information and influencing the opinion of other users in the network; which, in turn, can lead to trends and collective behaviour phenomena.

Rashid et al. [27] analyse social media for detecting masquerading or similarly deceptive behaviours. Using corpus linguistics and Natural Language Processing (NLP) techniques, such as Part-Of-Speech (POS) and semantic tagging, the authors predict key characteristics of individuals or groups hiding behind digital personas. POS tagging can provide information about the characteristics of the writing style, while semantic analysis allows classifying words based on their meaning, such as names, locations, emotional states and more. By applying these techniques, it is possible to detect any trends emerging during the conversation; for example, when a conversation is becoming increasingly aggressive.

Another example of what Language Style Matching (LSM) has to offer comes from Taylor et al. [33] who analyse the language changes in emails for detecting insider threats. In a series of simulations, the authors determined that insiders attempting to steal sensitive information did not change their overall work effectiveness (i.e., the strong signal), but they did show subtle changes in "extra-role" behaviours peripheral to their work. These included changes in their self-presentation to others and changes in their emotional engagement with colleagues (e.g., sympathising with work pressures). This leakage may be construed as a weak signal that reflects the psychological changes that occur prior to committing an insider offence against colleagues.

Conway et al. [11] analyse online news articles using a machine learning approach for detecting disease outbreaks. They show that a combination of words, n-grams and semantic features, in conjunction with feature selection techniques, can enhance the performance of a text classification model.

Other recent studies have demonstrated how social media content can be used for predicting real-world phenomena. Bollen et al. [7] analyse the mood of public tweets to predict changes in the stock market. Their results showed that an emotional change in tweets is followed by a similar rise or fall in the stock market prices from 2 to 6 days later. This indicates that mood (sentiment) changes in social media can possibly have predictive value regarding real-world phenomena. Sakaki et al. [31] propose an approach for detecting earthquakes in Twitter based on a probabilistic spatio-temporal model. The authors achieved to send email notifications about earthquakes much faster than the meteorological agencies. Sankaranarayanan et al. [32] present a system, called TwitterStand, that detects breaking news in Twitter using an online clustering algorithm, along with a text-based classifier for distinguishing *news* from *noisy* tweets. In [4], the authors show how time-series and sentiment analysis on tweets related to movies can actually predict box-office revenues. Achrekar et al. [2] present a framework, namely SNEFT, which predicts real-world flu epidemics through social media analysis. The authors show that the number of flu-related tweets is highly correlated with the rate of influenza cases reported by official statistics.

In contrast to the above studies, our work aims at using publicly available user-generated content to detect weak signals that could be early indicators of real-world collective actions, such as civil unrest. Additionally, as opposed to other works, our work is not limited to a single dimension (technique) in analysing the data, but considers a multi-dimensional approach consisting of various data analysis methods—both quantitative and qualitative.

3 Proposed Predictive Model

The proposed model (Fig. 7.1) consists of a variety of techniques for the detection of weak signals in social media. These techniques include: (1) keyword analysis, (2) geo-spatial analysis, (3) frequency analysis, (4) semantic analysis and (5) sentiment analysis. Using these techniques we aim to detect weak signals and their patterns that could lead to the prediction of future offline phenomena. For example, the detection of geo-location information (either in the form of coordinates or location names in text) in a higher frequency than in the past could be a form of a

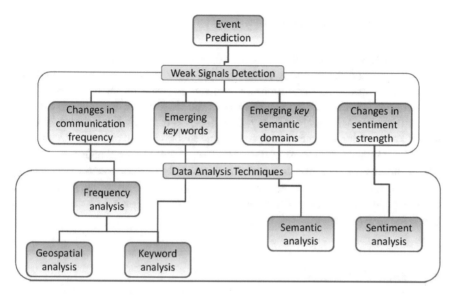

Fig. 7.1 Proposed predictive model

weak signal. Also, keyword and semantic analyses play a key role when it comes to detecting keywords and topics of interest. Semantic analysis allows the detection of further items that would be harder, if not impossible, to spot at the keywords level in such large-scale datasets. It groups together lower frequency words, as well as words that wouldn't be identified as key, and hence would not be observable otherwise. It also enables better insights into the concepts and topics discussed, hence, making it possible to identify emerging trends within a text much faster. A word, or semantic domain, is *key* if it occurs more often than would be expected by chance in a corpus compared to some "normative" corpus [28]. Keywords/semantics that initially appear at the fringes of online communications, but quickly gain strength over time, can be early indicators of future trends. Finally, sentiment analysis can significantly improve the prediction of events by giving insights into the emotions of users and help in detecting emotional shifts. The negative sentiment is usually the key to widespread events in Twitter; as such events are associated with rises in the negative sentiment [35]. Therefore, unusual rises in the negative sentiment could be an early indication of emerging trends. A more detailed description of the data analysis techniques is given in Sect. 4.3.

4 Evaluation Methodology

We evaluate our proposed model through analysis of tweets collected during the London riots in 2011.

4.1   The 2011 London Riots

Between 6 and 10 August 2011, riots took place across England in which buildings were burned down, shops looted, people were killed and property damages
amounted to hundreds of millions of pounds. A suspect's shooting[1] by police
on 4 August, at around 18:15, had sparked the riots which initially started as a
peaceful protest in Tottenham 2 days later (i.e., on 6 August), but within hours it
turned into massive violence and opportunistic looting.[2] Soon after, the riots spread
throughout London with some of the most affected areas being Woodgreen, Enfield
and Walthamstow. The use of social media not only helped rioters in organising
these events, but also caused more people to find out about the riots and get involved,
as well as inspired more violence in other locations too—what described by the
media as *copycat violence*.

4.2   Data Collection

Using Topsy,[3] we collected publicly available tweets posted between 4 and 10
August 2011 (totally, 831,041 tweets posted by 361,200 unique users). The rationale
behind collecting tweets—if any—2 days before the riots started is to observe
any potential changes in the emotions, behaviour, communication frequency, etc.;
particularly, after the shooting occurred. The most popular and trending topics
during the riots were selected, i.e., #tottenham and #londonriots. We queried Topsy's
database with combinations of keywords like: (1) #tottenham OR tottenham and
(2) #londonriots OR (#london AND riots), and used language-filtering to exclude
any non-English tweets. The restriction to English was selected to avoid any
complications of multiple languages in the results. Once tweets were downloaded,
we extracted additional information from them, such as user mentions, hashtags,
urls and timestamp (converted to GMT+1). From the collected tweets, we formed
two datasets, i.e., (1) *Tottenham* and (2) *London riots*.

4.3   Data Analysis

First, we use frequency analysis to identify the times at which communication
frequency changes and detect any potential correlations between tweets' rate and
the actual offline events. Further, we use geo-spatial analysis by extracting any
location information embedded in tweets (i.e., words referring to locations) and use

---

[1]http://theguardian.com/world/2011/aug/05/man-shot-police-london-arrest (Accessed: 3/3/2016).
[2]http://theguardian.com/uk/2011/aug/07/tottenham-riots-peaceful-protest (Accessed: 3/3/2016).
[3]http://en.wikipedia.org/wiki/Topsy_Labs (Accessed: 3/3/2016).

the Google charts API to generate heat maps of the initially online mentioned, but later actually affected locations. We also use SentiStrength,[4] a sentiment analysis tool, to detect any potential emotional changes in the communication before, during and after the incidents. SentiStrength is designed especially for estimating the sentiment in *short* texts (even for informal language) and can also deal with misspellings, emoticons, etc. It is thus a suitable tool for analysing the sentiment of *tweets*. SentiStrength classifies the text simultaneously as positive and negative by assigning a value for each text on a scale from $+/-1$ (*no* sentiment) to $+/-5$ (*very strong* positive/negative sentiment) [34]. Due to the high volume of tweets, we first classified all tweets by SentiStrength and then calculated the average positive and negative sentiment scores of all tweets per time-scale (e.g., per quarter-hour). We extend the analysis further by using natural language processing techniques to detect *key* words and topics/themes within the conversations. This analysis is aided by Wmatrix,[5] a corpus analysis and comparison tool. Wmatrix has been applied to numerous studies including political science research, online language analysis, corpus stylistics, as well as extremist language analysis [11, 24, 25, 27, 28]. Amongst other things, Wmatrix allows comparing a corpus (i.e., a set of texts) with a reference corpus (or another subset of the dataset) to obtain "key" items at different domain levels, such as *word* and *semantic*. In this paper, the *BNC Sampler Corpus Written*[6] was used as our reference corpus. The *BNC Sampler Corpus Written* is a sub-corpus of the British National Corpus, consisting of approximately one million words from a wide variety of written British English. It contains a wide and balanced sampling of texts from the BNC, maintaining the general text types (and their proportions) of the BNC as a whole. It is thus suited for comparing the collected English tweets with it. Wmatrix compares the relative frequencies of occurrence of items in the two corpora using the Log-Likelihood (LL) statistic [28]. The critical LL values (cut-off points) used by Wmatrix are the same as for the chi-squared distribution with 1 degree of freedom (d.f.), e.g., 3.84 (5% level), 6.63 (1% level), 10.83 (0.1% level) [29]. In the current analysis, items with an LL value of 6.63 or higher, i.e., at the 1% significance level (or $p < 0.01$), are considered to be "key" items (providing 99% certainty that the results are not due to chance). The most significant 100 key items are visualised as "tag cloud" in alphabetical order. The larger an item is, the greater its significance. Wmatrix also allows qualitative examination of the key items through the use of concordances by clicking on items within the cloud. This allows us to closely examine the context behind the key items and get more qualitative insights into users' thoughts, feelings and behaviour. Before undertaking the analysis with Wmatrix, tweets were converted to lowercase and cleansed of all information that would lead to mistaken results. More precisely, the cleaning process involved the removal of special characters, prefix "RT" (i.e., abbreviation of the word "re-tweet"), usernames and urls.

---

[4]http://sentistrength.wlv.ac.uk/ (Accessed: 3/3/2016).

[5]http://ucrel.lancs.ac.uk/wmatrix/ (Accessed: 3/3/2016).

[6]http://ucrel.lancs.ac.uk/bnc2sampler/sampler.htm (Accessed: 3/3/2016).

5   Results

5.1   The Tottenham Riots

5.1.1   The Shooting Incident

Figure 7.2a shows the hourly number of tweets from *Tottenham* dataset between 4 and 10 August 2011. As can be seen in Fig. 7.2a, although the flow of tweets was initially stable and low in frequency, however, in the evening of 4 August it was followed by a slight increase. Figure 7.3a, which depicts the hourly average sentiment of all tweets from Tottenham dataset, shows that there has been a sharp increase in the negative sentiment during that time-period, whereas, at the same time, the positive sentiment fell (−ve = −2.18 and +ve = 1.22 on 5 August at 01:00). The linguistic analysis of tweets, from 18:00 on 4 August to 01:00 on 5 August, shows that this weak, however, important rise on tweets' rate is strongly associated with the incident that occurred. Figure 7.4a illustrates the tag cloud with the most significant keywords as calculated by Wmatrix. Keywords such as *shooting* (LL = 743.38), *dead* (LL = 549.22), *police_officer* (LL = 450.59) and *police* (LL = 389.21) are among the top overused ones. Semantic analysis (Fig. 7.5a) also shows the categories *Warfare,_defence_and_the_army;_weapons* (LL = 846.32), *Dead* (LL = 381.59), *Law_and_order* (LL = 317.83), *Time:_Present;_simultaneous* (LL = 20.13) and *Vehicles_and_transport_on_land* (LL = 64.73) to be among the predominant key semantic domains. A concordance analysis of the content behind these key items (both words and semantics) shows clearly that the conversation revolves around the *incident* (LL = 142.98) and the *man* (LL = 547.80) who *got* (LL = 73.96) *shot* (LL = 779.70) at Tottenham *hale* (LL = 760.98). Specifically, one user posted: *"Arrrggghh The guy who got shot in Tottenham Hale died :("* (4 Aug 20:13). *"Another death in tottenham! Smh!!"* (4 Aug 22:57), wrote another. The emotional shift, along with the *key* words and themes, indicates weak signals of the growing collective anger towards the police; which would eventually become one of the major reasons for the riots.[7] These early signals could have given insights into the behaviour and emotions of people before the protest started on 6 August—which would shortly turn violent.

5.1.2   The Beginning of Protest

On 6 August, as shown from Fig. 7.2a, there has been a gradual increase from 18:00 to 20:00, while after 21:00 tweets' rate dramatically rose. Figures 7.4b and 7.5b show the most significant keywords and semantics between 18:00 and

---

[7]http://theguardian.com/uk/2011/dec/05/anger-police-fuelled-riots-study (Accessed: 3/3/2016).

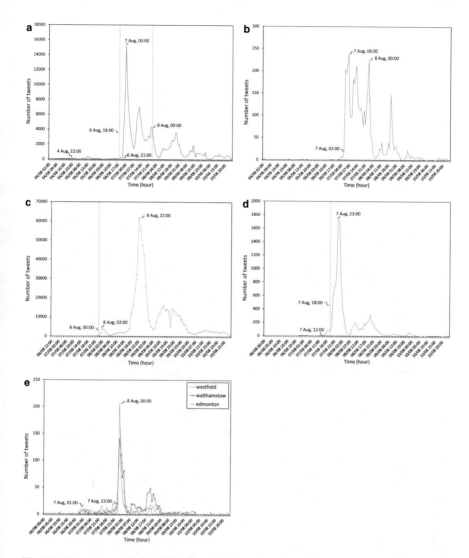

Fig. 7.2 Frequency analysis. (**a**) Number of *Tottenham* tweets. (**b**) Number of tweets containing *Woodgreen* or *Wood green*. (**c**) Number of *London riots* tweets. (**d**) Number of tweets containing *Enfield*. (**e**) Number of tweets containing *Walthamstow, Westfield and Edmonton*

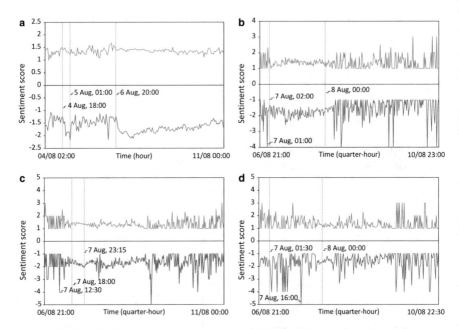

Fig. 7.3 Sentiment analysis. (**a**) Average sentiment of *Tottenham* tweets. (**b**) Average sentiment of tweets containing *Wood green* or *Woodgreen*. (**c**) Average sentiment of tweets containing *Enfield*. (**d**) Average sentiment of tweets containing *Walthamstow, Westfield* and *Edmonton*

20:00. At first glance, it seems that the whole discussion revolves around sports. However, examining the key items more closely, it can be seen that the discussion about the protest appears at the fringes of this mainstream topic. Content analysis of keywords shows that tweets refer to the *march* (LL = 47.15) that is *kicking_off* (LL = 42.49) *right_now* (LL = 52.51) outside the *police_station* (LL = 54.75) on Tottenham *high* (LL = 99.99) *road* (LL = 92.80). Key semantics (appearing at the periphery) include *Time:_Present;_simultaneous* (LL = 21.75), *Time:_Beginning* (LL = 17.62), *Law_and_order* (LL = 7.88), *Vehicles_and_transport_on_land* (LL = 29.20) and *Violent/Angry* (LL = 44.88), as can be seen in Fig. 7.5b. All these semantics indicate the beginning of the protest, as well as the likely ensuing violence. Tweets—many of them of aggressive and hostile nature—encourage others to join in. For instance, one user posted: *"If you hate police come Tottenham high road now"* (6 Aug 19:05). These early signals show that the discussion around the protest was becoming an emerging topic. Also, they give insights regarding the hostility of some people towards the police and their efforts to mobilise the public.

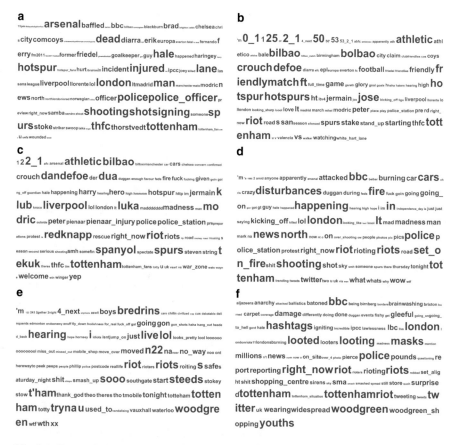

Fig. 7.4 Keyword clouds. (**a**) *Tottenham* (4 Aug 18:00–5 Aug 01:00). (**b**) *Tottenham* (6 Aug 18:00–20:00). (**c**) *Tottenham* (6 Aug 20:00–21:00). (**d**) *Tottenham* (6 Aug 21:00–22:00). (**e**) *Woodgreen* (6 Aug 20:00–7 Aug 02:00). (**f**) *Woodgreen* (7 Aug 05:00–7 Aug 06:00)

5.1.3 The Protest Turns Violent

As shown in Fig. 7.3a, after 20:00 on 6 August, the negative sentiment sharply increased while the positive sentiment slightly dropped. Analysing the tweets from 20:00 to 21:00 on 6 August (Fig. 7.4c), it appears that keywords such as *police_station* (LL = 83.65), *right_now* (LL = 101.29) and *riot* (LL = 281.43) appear significantly stronger than in the previous hours. Tweets refer to the group of people who attacked and set on *fire* (LL = 58.79) two *police*

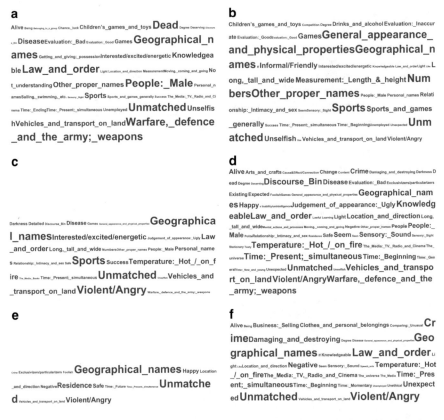

Fig. 7.5 Key semantic clouds. (**a**) *Tottenham* (4 Aug 18:00–5 Aug 01:00). (**b**) *Tottenham* (6 Aug 18:00–20:00). (**c**) *Tottenham* (6 Aug 20:00–21:00). (**d**) *Tottenham* (6 Aug 21:00–22:00). (**e**) *Woodgreen* (6 Aug 20:00–7 Aug 02:00). (**f**) *Woodgreen* (7 Aug 05:00–7 Aug 06:00)

(LL = 143.48) *cars* (LL = 68.22) (i.e., the time that the peaceful protest turned violent). Key semantic tags (Fig. 7.5c) include *Law_and_order* (LL = 56.64), *Time:_Present;_simultaneous* (LL = 11.11), *Temperature:_Hot_/_on_fire* (LL = 68.21), *Vehicles_and_transport_on_land* (LL = 66.15), *Warfare,_defence_and_the_army;_weapons* (LL = 9.72) and *Violent/Angry* (LL = 107.87). All these semantic domains highlight the situation predominated that time outside the Tottenham's police station. Although the discussion about the riots was still at the fringes, it was increasingly emerging and would shortly pass to the mainstream.

5.1.4 Riots Spread Across Tottenham

As can be seen from Fig. 7.2a, the volume of tweets rapidly climbed after 21:00 on 6 August and reached a peak by the end of the day. Figure 7.4d, which shows the keyword analysis between 21:00 and 22:00 on 6 August, reveals that the conversation about the *disturbances* (LL = 1061.44) outside the *police_station* (LL = 557.28) had moved from the fringes and became the trending topic on Twitter. Keywords such as *police* (LL = 6350.46), *riots* (LL = 3209.72), *shooting* (LL = 1023.18) and *right_now* (LL = 963.56) are among the top significant keywords. From the semantic analysis of this time-period (Fig. 7.5d), it can be clearly seen that the main theme is the on-going riots. Key semantic domains include *Violent/Angry* (LL = 6194.84), *Warfare,_defence_and_the_army;_weapons* (LL = 443.39), *Temperature:_Hot_/_on_fire* (LL = 2903.46), *Crime* (LL = 148.05), *Law_and_order* (LL = 3936.67), *Vehicles_and_transport_on_land* (LL = 3175.43), *Time:_Beginning* (LL = 122.79) and *Time:_Present;_simultaneous* (LL = 698.89). The content analysis shows that people not only distributed the news, but also incited the crowd to violence and looting. Examples include: (1) 6 Aug 20:58—*"Who wants to come loot with me in tottenham?"* and (2) 6 Aug 22:24—*"En route to Tottenham to break into the jewellery shop"*. These signals could have given insights into the escalation of the situation that would follow in the next hours, as well as the possible locations of future attacks by rioters.

5.1.5 Riots Spread to Woodgreen

In the early hours of 7 August, the nearby Woodgreen was affected by riots; whereas police was still not present at 04:00.[8] As can be seen in Fig. 7.2b, from 20:00 on 6 August to 02:00 on 7 August, there has been an unusual growth in the number of tweets. The sentiment analysis of this time-period (Fig. 7.3b) shows a steep rise in the negative sentiment, reaching a score of −4 on 7 August at 01:00. These spikes in the negative sentiment indicate a possible emerging event. Before uploading the texts to Wmatrix, occurrences of *"Wood green"* were converted to *"Woodgreen"* to avoid any mistaken results in the analysis. Analysing the tweets from 20:00 on 6 August to 02:00 on 7 August (Fig. 7.4e), it appears that users warn that *rioters* (LL = 14.45) have *moved* (LL = 22.44) from Tottenham and they are *tryna* (LL = 28.90) *start* (LL = 17.14) a *riot* (LL = 102.53) in Woodgreen *N22* (LL = 28.90). Moreover, they mention incidents they eye-witness, such as that people *smash_up* (LL = 14.45) the *t-mobile* (LL = 14.45) shop. Semantic analysis (Fig. 7.5e) shows categories such as *Location_and_direction* (LL = 7.97), *Crime* (LL = 6.68), *Time:_Future* (LL = 7.86), *Time:_Present;_simultaneous* (LL = 6.78) and *Violent/Angry* (LL = 48.04) being among the emerging themes.

---

[8]http://theguardian.com/uk/2011/aug/07/tottenham-riot-looting-north-london
(Accessed: 3/3/2016).

These emerging key items (words and semantics) give early warnings of the impending riots in Woodgreen. After 02:00 on 7 August, tweets' rate sharply increased and reached a peak by 06:00, as can be seen in Fig. 7.2b. As shown in Fig. 7.3b, the negative sentiment consistently outweighed the positive sentiment in the following hours, giving insights into the current emotions of people. Keyword analysis from 05:00 to 06:00 on 7 August (Fig. 7.4f) shows that *riots* (LL = 541.06) had moved to Woodgreen and become a trending topic. Specifically, people report that the *shopping_centre* (LL = 57.69) is being *looted* (LL = 188.91) *right_now* (LL = 165.72) by *youths* (LL = 152.07) wearing *masks* (LL = 125.94). The semantic analysis (Fig. 7.5f) shows the emergence of the previously peripheral key domains such as *Crime* (LL = 212.42), *Time:_Present;_simultaneous* (LL = 66.69) and *Violent/Angry* (LL = 363.15). In addition, among the top key concepts are *Law_and_order* (LL = 120.43), *Time:_Beginning* (LL = 33.04), *Tempera-ture:_Hot_/_on_fire* (LL = 62.46), *Damaging_and_destroying* (LL = 47.26) and *Vehicles_and_transport_on_land* (LL = 13.15).

5.2   The London Riots

While the riots in Tottenham were fading out, disturbances would start spreading across other areas of London (Fig. 7.6b); with the situation being escalated in the evening of 8 August (Fig. 7.6e). As can be seen from Fig. 7.2a, the volume of *Tottenham* tweets quickly went down after 00:00 on 8 August. However, keyword analysis of *Tottenham* tweets after 00:00 on 8 August reveals the gradual rise of the keyword *londonriots* at the times: 00:00 (LL = 140.09), 01:00 (LL = 368.76) and 02:00 (LL = 1263.35). This early indication could have given warnings about the following shift from *Tottenham* to *London* riots and the impending massive violence throughout London. As can be seen in Fig. 7.2c, there has been a sudden rise in the number of tweets (from *London riots* dataset) between 00:00 and 06:00 on 8 August. The *londonriots* keyword was accompanied by names of various London boroughs that were about to get affected by riots in the following hours. Some of these areas were Enfield, Walthamstow, Westfield and Edmonton.

5.2.1   Riots Spread to Enfield

Figure 7.2d shows the hourly frequency of tweets containing Enfield. As shown in the figure, between 12:00 and 18:00 on 7 August, the number of tweets suddenly went up. This increase in frequency was accompanied by a rise in the negative sen-timent too, as can be seen in Fig. 7.3c. Keyword analysis in Fig. 7.7a shows people on twitter spreading *rumours* (LL = 162.46) saying that *riots* (LL = 588.21) are due to *start* (LL = 68.59) in Enfield at *4pm* (LL = 190.18). Others hear that more *trouble* (LL = 53.69) is expected in Enfield *tonight* (LL = 61.68). Key semantic

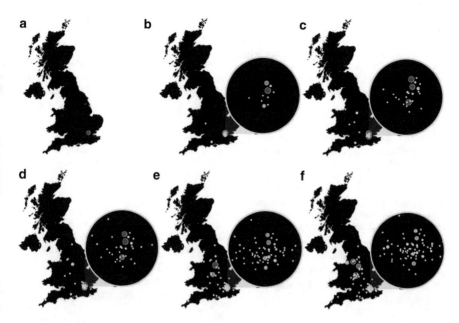

Fig. 7.6 Locations mentioned in tweets and which actually affected by riots. (**a**) 7 Aug at 00:00.
(**b**) 8 Aug at 00:00. (**c**) 8 Aug at 06:00. (**d**) 8 Aug at 16:00. (**e**) 8 Aug at 22:00. (**f**) 11 Aug at 00:00

categories (Fig. 7.7d) include *Crime* (LL = 45.80), *Damaging_and_destroying*
(LL = 31.02), *Law_and_order* (LL = 111.23) *Time:_Present;_simultaneous*
(LL = 118.08) and *Violent/Angry* (LL = 484.65).

As can be seen from Fig. 7.2d, the volume of tweets considerably rose after
19:00 on 7 August and reached a peak by 23:00. Figure 7.7b, which shows the
keyword analysis from 22:00 to 23:00 on 7 August, indicates that riots had moved
to Enfield. People report that *protestors* (LL = 914.32) are *throwing* (LL =
1057.91) *petrol_bombs* (LL = 1098.93) on *passing* (LL = 911.51) *cars* (LL =
905.48). Others describe incidents unfolding next to them, such as that *looters*
(LL = 173.62) are *looting* (LL = 732.34) the *tesco* (LL = 72.34) and *krispy
kreme* (LL = 376.18) shops. Semantic analysis (Fig. 7.7e) shows domains such
as *Violent/Angry* (LL = 1533.72), *Time:_Present;_simultaneous* (LL = 264.69),
Crime (LL = 1588.14), *Vehicles_and_transport_on_land* (LL = 458.30), *Damag-
ing_and_destroying* (LL = 86.54) and *Warfare,_defence_and_the_army;_weapons*
(LL = 252.86) being the prevalent themes.

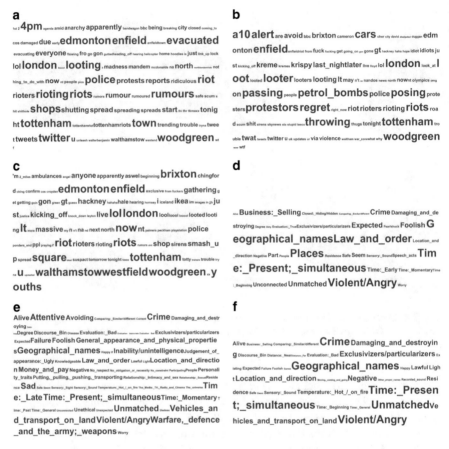

Fig. 7.7 Keyword and semantic clouds. (**a**) *Enfield* (7 Aug 12:00–18:00). (**b**) *Enfield* (7 Aug 22:00–23:00). (**c**) *Walthamstow, Westfield, Edmonton* (6 Aug 21:00–7 Aug 23:00). (**d**) *Enfield* (7 Aug 12:00–18:00). (**e**) *Enfield* (7 Aug 22:00–23:00). (**f**) *Walthamstow, Westfield, Edmonton* (6 Aug 21:00–7 Aug 23:00)

5.2.2 Riots Spread to Walthamstow, Westfield and Edmonton

Figure 7.2e shows the rate of tweets containing Walthamstow, Westfield and Edmonton. As shown in the figure, in the early morning of 7 August, these areas suddenly started emerging in the conversation on Twitter; whereas, after 23:00, the volume of tweets rapidly increased. The sentiment analysis (Fig. 7.3d) shows a similar pattern (i.e., sudden spikes in the negative sentiment) to the previously analysed boroughs before the event escalated. As can be seen in Fig. 7.3d, the negative sentiment sharply increased after 01:30 on 7 August, reaching high negative scores. Keyword analysis (Fig. 7.7c), from 21:00 on 6 August to 23:00 on 7 August, shows that people are *hearing* (LL = 41.19) that Westfield is going to be hit by riots. One wrote, *"...these broadcasts telling me to meet up to trash westfield"* (7 Aug 14:46). Others report that they are *getting* (LL = 40.28) messages saying that riots are going to kick off in Walthamstow. In Edmonton, people report that the *jjb* (LL = 30.78) shop is being *looted* (LL = 61.57) while *police* (LL = 78.74) is dealing with riots in Tottenham. Shortly after 22:00 on 7 August, people report that a *massive* (LL = 67.43) group of *youths* (LL = 119.29) is *gathering* (LL = 78.40) in Walthamstow *beginning* (LL = 42.75) to *smash_up* (LL = 112.87) a *shop* (LL = 94.51). Semantic analysis (Fig. 7.7f) shows categories such as *Crime* (LL = 98.47), *Damaging_and_destroying* (LL = 38.03), *Vehicles_and_transport_on_land* (LL = 29.45), *Time:_Present;_simultaneous* (LL = 113.09), *Time:_Beginning* (LL = 14.26) and *Violent/Angry* (LL = 335.16) being amongst the top key topics. These weak signals could clearly provide early warnings about the beginning and the impending escalation of riots throughout these areas in the following hours.

6 Machine Learning Model

Using the insights from the above analysis, we proceeded to build a machine learning model that would be predictive of violent crowd behaviour and collective action in social media. This task is aided by the *Weka*[9] toolkit which provides a collection of machine learning algorithms and data pre-preprocessing tools [18]. Weka allows users to import an ARFF (Attribute-Relation File Format) file with a list of instances sharing a set of attributes (or features), which can be used to train or test a classifier.

6.1 Data Labelling

For this task, we have selected a representative sample of tweets (excluding duplicates or re-tweets) from the beginning of *Tottenham riots* dataset to the first violent outbreak on 6 August, as well as early tweets (i.e., before tweets' rate reaching its peak) from subsequent affected locations analysed in this paper. These tweets were manually labelled with the best matching class, i.e., *Relevant* or *Non-Relevant*. Tweets referring to the shooting incident or the beginning of protest/riots, showing anger towards the police, being supportive and inciting violence, would be labelled as *Relevant*. Tweets with no any of these characteristics would be labelled as *Non-Relevant*. Out of 6011 tweets, 3009 were labelled as *Relevant* and 3002 were labelled as *Non-Relevant*.

---

[9]http://www.cs.waikato.ac.nz/ml/weka (Accessed: 3/3/2016).

6.2   Feature Extraction

The feature vector consists of *key* words, semantics, Part-of-Speech (POS) tags, positive and negative sentiment. Statistically significant items at word, semantic and POS level were obtained by comparing the collection of tweets (after being cleansed and pre-processed) labelled as *Relevant* with the BNC Sampler Corpus Written using Wmatrix. By setting the Log-Likelihood cut-off at 6.63, 928 keywords and Multi-Word Expressions (MWEs), 56 semantic domains and 37 POS tags were retained (= 1023 features in total—including positive and negative sentiment). We opted to exclude stemming and stop-words removal as stemming could undermine subsequent feature selection by conflating and neutralising discriminative features, while stop-words could be highly discriminative features for text classification, as found by Yu [38]. Additionally, overly common words (including stop-words) that are unlikely to aid classification are scaled down at the *keyness* analysis, i.e., when the collection of tweets is compared with the reference corpus.

6.3   Data Preparation

Using Wmatrix's API, all 6011 tweets were individually tagged and stored in a database. Wmatrix's server responds by returning an XML document including the tweet in a tokenised form. Each *token* is accompanied by a POS and semantic tag (which correspond to a *syntactic/grammatical category*[10] and *semantic domain*[11], respectively) as identified by the CLAWS [15] and USAS [30] systems. The response document also provides indication for any MWEs as identified by Wmatrix [23], which helps us to automatically extract any multi-word units in tweets, such as *kicking_off* and *right_now*. An example of a tweet tagged by Wmatrix can be seen in Fig. 7.8. Using Java and Weka's API, the tagged tweets were retrieved from the database to be transformed into a list of instances. The presence or absence of each feature (i.e., features that extracted in the *Feature Extraction* phase) in tweet is treated as a boolean attribute (i.e., 1 for presence or 0 for absence). In short texts, such as tweets, words are unlikely to repeat, making boolean attributes nearly as informative as other representation methods (e.g., *term frequency*) [13]. Positive and negative sentiment scores are treated as numeric attributes as calculated by SentiStrength. This process resulted to an ARFF file that forms the dataset for training the classification model.

---

[10]http://ucrel.lancs.ac.uk/claws6tags.html (Accessed: 3/3/2016).
[11]http://ucrel.lancs.ac.uk/usas/USASSemanticTagset.pdf (Accessed: 3/3/2016).

```
<tweet>
<w lemma="there" pos="EX" sem="Z5">There</w>
<w lemma="be" pos="VBZ" sem="A3+">'s</w>
<w lemma="a" pos="AT1" sem="Z5">a</w>
<w lemma="riot" pos="NN1" sem="E3-">riot</w>
<w lemma="in" pos="II" sem="Z5">in</w>
<w lemma="tottenham" pos="NP1" sem="Z2">Tottenham</w>
<w lemma="PUNC" pos="!" sem="">!</w>
<w lemma="it" pos="PPH1" sem="Z8">It</w>
<w lemma="be" pos="VBZ" sem="A3+">'s</w>
<w lemma="all" pos="DB" sem="N5.1+">all</w>
<w lemma="kick" pos="VVG" sem="T2+[i1.2.1">kicking</w>
<w lemma="off" pos="RP" sem="T2+[i1.2.2">off</w>
<w lemma="on" pos="II" sem="Z5">on</w>
<w lemma="the" pos="AT" sem="Z5">the</w>
<w lemma="high" pos="JJ" sem="N3.7+">high</w>
<w lemma="rd" pos="NN1" sem="Z2">rd</w>
<w lemma="right" pos="RR" sem="T1.1.2[i2.2.1">right</w>
<w lemma="now" pos="RT" sem="T1.1.2[i2.2.2">now</w>
<w lemma="PUNC" pos="!" sem="">!</w>
</tweet>
```

Fig. 7.8 Example of tweet tagged by Wmatrix

6.4 Feature Selection

Although many unhelpful words, semantics and POS tags were eliminated in the *Feature Extraction* phase, an overwhelming number of features still remain. It is common to use feature selection (also known as attribute selection) algorithms to remove irrelevant or redundant attributes that could confuse the machine learning algorithms [36]. Such methods not only can reduce over-fitting and improve the performance, but they can also reduce the run-time of the learning algorithm (since the number of attributes is minimised). A well-known feature selection technique is *Information Gain* which is used to select the most informative/useful features that discriminate between the classes. Information Gain has been proven to be effective in the context of text classification [37]. It is thus a suitable method to use. Using Weka's *InfoGainAttributeEval* as the Attribute Evaluator and *Ranker* as the search method, we obtained a sorted list of attributes ranked according to their information gain score. This task was performed through Weka's meta-classifier, called *AttributeSelectedClassifier*, which selects attributes based on the training set only (even if cross-validation is used for testing the classifier), and then it trains the classifier again on the training set only [36]. Table 7.1 shows the 10 most discriminative attributes (representing the top 1% of the features) rank-ordered by Information Gain. As can be seen, the *Negative Sentiment* attribute highly contributed to making classification decisions. Then comes the *Violent/Angry (E3-)* and *Law_and_order (G2.1)* semantic domains. Keywords such as *riots, police* and *riot* are among the

Table 7.1 Top 10 attributes ranked by information gain

Rank	Attribute	Information gain
1	Negative sentiment	0.4657
2	E3-	0.26942
3	G2.1	0.11131
4	Riots	0.10216
5	Police	0.09644
6	Riot	0.09375
7	O4.6+	0.07819
8	II	0.056
9	G3	0.05245
10	M3	0.047

most discriminative features too. In the 7th place is the *Temperature:_Hot__on_fire (O4.6+)* semantic domain, followed by the *General preposition (II)* POS tag. General preposition is a category of lexical items that expresses spatial or temporal relationships, such as *on the streets*, *outside*, *near* and *at the moment*. In 9th and 10th place are the semantic domains *Warfare,_defence_and_the_army;_weapons (G3)* and *Vehicles_and_transport_on_land (M3)*, respectively. Of the 102 (i.e., top 10%) most discriminative features, 63 were keywords and MWEs, 21 were semantic tags, 16 were POS tags and 2 were sentiment scores (i.e., negative and positive sentiment). This implies that a combination of a variety of features could be an optimal approach to text classification, rather than relying on a single feature type.

6.5 Experiments and Results

Rather than selecting an arbitrary cut-off point in the feature selection process, we perform experiments with different number of the ranked features, i.e., top 1%, 5%, 10% etc., using a variety of learning algorithms, i.e., *Naive Bayes, SVM (using the third-party LIBSVM library [9]), J48 (C4.5 decision tree)* and *RandomForest*. We then investigate which subset of attributes, along with which learning algorithm, produces the best classification performance.

The performance of the classifiers is evaluated using *stratified* tenfold cross-validation. The dataset is divided into 10 subsets in which the *class* is represented in approximately the same proportions as in the whole dataset [36]. One of the 10 subsets is used as the test set and the remaining nine-tenths are used for training the learning algorithm. This process is repeated 10 times, with each of the 10 subsets used as the test set exactly once. Standard performance measures—*Recall, Precision* and *F-measure*—are used. These measures are defined as follows:

$$Recall = \frac{tp}{tp + fn} \tag{7.1}$$

$$Precision = \frac{tp}{tp + fp} \tag{7.2}$$

$$F - measure = 2 \cdot \frac{Recall \cdot Precision}{Recall + Precision} \tag{7.3}$$

The *true positives* (tp) and *true negatives* (tn) are the numbers of correctly classified instances, while *false positives* (fp) and *false negatives* (fn) are the numbers of incorrectly predicted outcomes. Recall is the proportion of instances which were classified as positive, among all instances which are actually positive. Precision is the proportion of instances which are actually positive, among all those which were classified as positive. The F-measure is the harmonic mean of precision and recall. As we aim to investigate the relevance classification performance, we focus on the precision, recall and F-measure of the *Relevant* class (i.e., the positive class).

6.5.1   Naive Bayes

Table 7.2 shows the performance results of Naive Bayes classifier when selecting different number of the most discriminative features each time. As can be seen in the table, precision is relatively stable (at around 0.915) when selecting 5% or more of the features, while recall improves as more features are added. The best performance is achieved when the top 75% of the features is used. This results in a 0.921 precision, 0.905 recall and 0.913 F-measure.

6.5.2   SVM

Table 7.3 shows the performance results of SVM classifier. As shown in the table, the highest recall is achieved when only 1% of the features is used, however, this results in a rather low precision. By using the top 10% of the features, the precision jumps to 0.909 with a fairly high recall of 0.908 and F-measure of 0.909.

Table 7.2 Naive Bayes results

Features selected (%)	Precision	Recall	F-measure
1	0.902	0.856	0.878
5	0.915	0.879	0.896
10	0.915	0.89	0.902
15	0.915	0.894	0.904
25	0.915	0.902	0.908
50	0.919	0.905	0.912
75	0.921	0.905	0.913
100	0.92	0.904	0.912

Table 7.3 SVM results

Features selected (%)	Precision	Recall	F-measure
1	0.854	0.916	0.884
5	0.908	0.909	0.908
10	0.909	0.908	0.909
15	0.907	0.91	0.908
25	0.902	0.913	0.907
50	0.895	0.913	0.904
75	0.895	0.914	0.904
100	0.895	0.912	0.903

Table 7.4 J48 results

Features selected (%)	Precision	Recall	F-measure
1	0.897	0.865	0.881
5	0.91	0.899	0.905
10	0.922	0.897	0.909
15	0.924	0.896	0.91
25	0.926	0.892	0.908
50	0.934	0.883	0.908
75	0.934	0.881	0.907
100	0.936	0.884	0.909

6.5.3 J48

As opposed to the previous two classifiers, J48 classifier seems to be increasingly achieving higher precision as the number of features increases; reaching a precision of 0.936 when all features are used (Table 7.4). However, this has an impact on the recall which drops as the precision increases. The classifier yields the best performance when the top 15% of the features is used. This results in a 0.924 precision, 0.896 recall and 0.91 F-measure.

6.5.4 RandomForest

RandomForest is the classifier that outperforms others with a 0.945 precision, 0.903 recall and 0.923 F-measure (when 100% of the features is used). As can be seen in Table 7.5, RandomForest can also achieve fairly high performance with even smaller number of features used, maintaining high precision and recall levels.

6.6 Model Selection

It is important to select a model that can achieve sufficiently high performance with the least possible number of features (if such a classifier is to be used for

Table 7.5 RandomForest results

Features selected (%)	Precision	Recall	F-measure
1	0.901	0.856	0.878
5	0.909	0.907	0.908
10	0.922	0.912	0.917
15	0.93	0.911	0.92
25	0.937	0.903	0.92
50	0.943	0.899	0.921
75	0.944	0.895	0.919
100	0.945	0.903	0.923

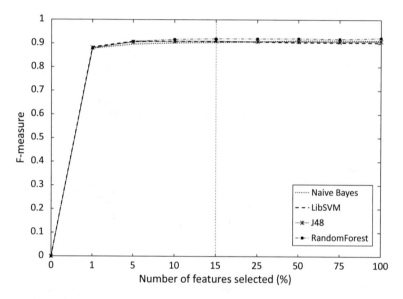

Fig. 7.9 Effect of feature set size on F-measure

real-time classification). Figure 7.9 depicts the effect of feature set size (i.e., number of features selected) on the F-measure for each of the classifiers.

As Fig. 7.9 shows, there is a rise in the performance of all classifiers between 1% and 15% of features selected, with *RandomForest* being the predominant classifier. The performance of all classifiers begins to stabilise at 15% of features selected. This suggests that selecting a classifier which uses more than 15% of the features would not bring about any significant improvement on the performance of the classifier, but, on the contrary, would cause a decrease on the recall, as can be seen in Fig. 7.10. Figure 7.10 shows that, as the feature set size increases, the recall (after 10% of features selected) considerably drops for classifiers *J48* and *RandomForest*. In the case of *SVM* and *Naive Bayes* classifiers, the recall remains relatively stable, or slightly increases without any notable changes.

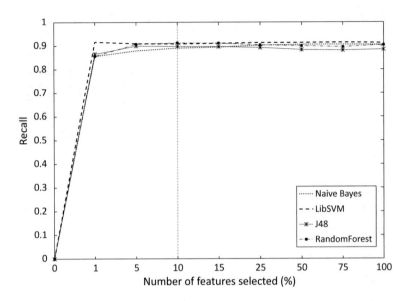

Fig. 7.10 Effect of feature set size on Recall

Table 7.6 Selected classifier detailed results

Class	TP Rate	FP Rate	Precision	Recall	F-measure
Relevant	0.912	0.077	0.922	0.912	0.917
Non-relevant	0.923	0.088	0.913	0.923	0.918
(Weighted avg.)	0.917	0.083	0.918	0.917	0.917

In this particular classification problem, it is desirable for the classifier to achieve the highest possible recall (by using a relatively small number of attributes), while maintaining precision at acceptable levels. In other words, it is important to identify as many relevant tweets (i.e., true positives) as possible, by minimising the number of tweets that would be incorrectly classified as non-relevant (i.e., false negatives), but are actually relevant. The classifier that meets this goal is RandomForest when the feature set consists of the top 10% of the attributes. This classifier results in a 0.922 precision, 0.912 recall and 0.917 F-measure, as shown in Table 7.5. For the reasons outlined above, we have selected this particular classifier to be designated as the model for future real-time classification of micro-blog data. Table 7.6 presents the detailed results (including true positive and false negative rates) of the selected classifier, as well as the weighted average score. The true positive rate is defined in the same way as Eq. (7.1), while false positive rate is defined as follows:

$$FPR = \frac{fp}{fp + tn} \qquad (7.4)$$

7 Discussion

From our analysis, we have derived predictive patterns, such as changes in communication frequency, emotion and language, that could constitute the weak signals in social media. Each of these weak signals could act as a weak signal on its own, but also, its coexistence along with other weak signals could form a "stronger" (or, of higher importance) weak signal. In other words, if a weak signal is accompanied by other weak signals too, then the importance of this weak signal is increased. The higher the combination of weak signals, the greater their importance. The importance level of weak signals plays a key role in the early and reliable prediction of future events, since a weak signal on its own could be overlooked due to the high volume of data in these networks, or could be vague and irrelevant. However, if a cluster of weak signals—signals that are related to each other—is detected, the possibility of these weak signals becoming a trend is much higher. Below, we provide examples of the weak signals identified by the analysis conducted in this paper.

The first weak signal observed in all cases that we examined was the unusual, and often steep, rise in the frequency of communication. The rate of messages on Twitter seems to follow a path that is closely related to emerging events in the real-world, as shown by the frequency analysis. This unusual increase in frequency, compared to the normal rate, is a weak signal of a possible emerging trend. Our findings also show that such a sudden rise in the volume of messages is usually accompanied by spikes in the negative sentiment. Therefore, a sharp increase in the negative sentiment is another potential weak signal. This combination of increase in the frequency and negative sentiment of tweets could form a "stronger" weak signal, with the probability that an event is emerging to be higher.

It is also noticeable how people use Twitter to instantly report what is unfolding in front of their eyes and being specific by giving locations. It is also interesting the way they share broadcast messages from other social networking services in an effort to warn the general public for upcoming disorders in specific locations. As shown by the geo-spatial analysis, a number of locations that were about to be affected by riots were first reported by users on Twitter. The analysis of messages containing some of these areas showed that the detection of weak signals, indicating the forthcoming disorders, was possible. Furthermore, the depiction of affected locations with the use of heat maps shows the riots to begin from Tottenham, quickly spread to adjacent areas, and eventually, followed by similar disorders in other parts of London and England. Consequently, a geo-spatial analysis, with the help of heat maps, proves to be valuable in the case of such phenomena. Heat maps can reveal areas of increasingly higher activity that would allow the early and adequate deployment of police forces to prevent any possible forthcoming violent outbreaks.

An even more important contribution to the detection of weak signals was the *keyness* analysis at word and semantic levels. As shown from both of these analyses, keywords and semantic concepts that initially appear at the fringes of the communication, but increasingly gain strength (their significance, or Log-Likelihood value, increases) over time, provide early indications of emerging events. Therefore, the appearance of key items relevant to the context of interest (in this paper, that is, civil unrest), but more importantly, the increasing significance of these items over time is an important weak signal. For instance, the *Violent/Angry* domain seems to be one of the emerging key themes in such phenomena that increasingly becomes stronger over time. This semantic domain is associated with words and phrases that indicate violent actions and aggressive behaviours such as *riot, disturbances, fight, kick* and *attack*. If this domain or words are emerging within the online conversations, they could be early indicators of impending disturbances and violent actions. Similarly, the *Crime* domain, which is associated with criminal behaviour such as *looting, stealing* and *breaking into*, as well as the *Damaging_and_destroying* domain which is connected to *breaking, smashing_up, destroying*, etc. In addition, *Law_and_order* was one of the predominant key themes, since *police* is usually directly related to such events. Thus, the emergence of this theme can give early indications of possible aggressive behaviours towards the police. *Geographical_names* is also one of the key domains that could be used in conjunction with geo-spatial analysis, as this semantic domain includes information related to locations and places. The detection of *key* geographical names and places can provide weak signals of emerging local events. Moreover, the *Vehicles_and_transport_on_land* domain, which includes terms such as *vehicles, cars, streets* and *road*, seems to be a key topic in such phenomena; since these collective actions take place in the streets and often result in vandalising behaviours such as burning vehicles. Last but not least, *Time:_Present;_simultaneous* and *Time:_Beginning* domains can provide early warning signals about events happening *right_now, today, tonight*, or events that are *starting, kicking off, beginning*, etc. Both of these domains, along with the words and MWEs they enclose, are of great importance, as they can provide temporal indications, i.e., indications of time an event is going to take place. Therefore, the existence or absence of these key items in the conversation provides insights into whether the conversation refers to a forthcoming or emerging event, or to an event that has already passed.

The emergence of each key item mentioned above could indicate a weak signal. However, a combination of these weak signals could form a weak signal of higher importance. Further, if such a combination is accompanied by an unusual growth in the frequency of tweets, as well as a sudden rise in the negative sentiment, would add more value to the importance of this weak signal—and vice versa. For example, as shown by the frequency analysis of Tottenham tweets during the time-periods when (1) the shooting incident occurred and (2) the protest began, the change in communication frequency was not as steep as in other areas analysed; and therefore, such a signal could be overlooked by an approach that relies only on frequency analysis. However, the sudden rise in the negative sentiment, during both of these time-periods, strengthens the importance of these weak signals. Finally, the keyword and semantic analyses further increase the importance of these signals by revealing emerging words and topics, as well as giving insights into discussions unfolding online, thus making the early detection of these emerging changes possible.

Weak signals can be used as a tool to anticipate future changes/events and take appropriate and timely action. The proposed model could possibly be applied for detecting weak signals in domains other than civil unrest; since the keyword and semantic analyses are not domain restricted.

Motivated by the patterns identified in the previous analysis, we applied a machine learning approach to train a classification model that would be able to identify tweets related to civil unrest, with a view to predicting such mass phenomena. From the experiments conducted, a classifier based on the RandomForest learning algorithm, using only the top 102 (10%) out of 1023 attributes, was selected as the best performing classifier. While it was possible to select a classifier with higher precision, and hence with higher F-measure, it was, however, preferable to keep the true positive rate at the highest possible level. Nevertheless, with the selected classifier, precision still remains high, maintaining in this way a low false positive (also known as false alarm) rate.

One of the limitations of this study was the lack of coordinates in our datasets from any possible geo-tagged tweets. Although the amount of authors having enabled the geo-tagging feature would normally be small, we could possibly get more precise information as to where some of these tweets were coming from while the riots were forming. Also, if the collection of tweets containing the affected areas was performed as a separate query on Topsy, the detection of weak signals could significantly be improved—since in this analysis the names of these areas had to co-occur with the keywords we queried Topsy's database.

8 Conclusion and Future Work

Undoubtedly, social media facilitate the organisation of events and interaction among masses of people in a way that no other communication tool did before. They increase the possibility for leaders to exercise their influence on others, as well as give voice to introverts, i.e., people that would normally stay silent. More importantly, Twitter seems to act as a valuable reporting tool for anyone who can instantly report and give information from their position.

In this paper, we analysed Twitter data from the London riots in 2011 using a multi-disciplinary approach in an attempt to identify and detect weak signals to predict tipping points in that context. Our approach consisted of using various data analysis techniques, namely frequency, keyword, geo-spatial, semantic and sentiment analysis, as well as training a machine learning model for tweets classification. The results demonstrate that weak signals indeed exist in social media and that the proposed model can be successfully used in the detection of these early indicators. Also, linguistic, semantic and sentiment features, combined with feature selection techniques such as Information Gain, have been found to be effective predictors in text classification. The evaluation of the classification model reports encouraging results for the prediction of such real-world phenomena. The selected classifier could be applied in combination with the data analysis techniques to scale up the weak signals detection. The detection of such predictive signals in social media could inform law enforcement agencies to take precautions, as well as countermeasures, in the case of impending violent collective actions.

In future, we aim to examine the usefulness of social network analysis techniques in identifying potentially predictive behaviours hidden in the connections between individuals and groups. We will study the importance of these connections and actors (both central and peripheral) in spreading the information in the network and influencing their neighbours. We also aim to use collocation analysis techniques (provided by Wmatrix, at the word and word-semantic levels) to examine networks and links between words, actors and geographical places, etc., to identify possible weak signals and their patterns for forecasting future events. Furthermore, we will explore additional features derived from the above analyses (i.e., social network analysis and collocation analysis) and their usefulness in enhancing the performance of the classification model. Finally, we will use text normalisation tools, such as VARD [5], as a pre-processor to NLP techniques (e.g., keyword analysis, POS and semantic tagging), with the aim to reduce the amount of spelling mistakes/variants in tweets—which will improve the accuracy of these techniques, and hence, the accuracy of the classification model.

Acknowledgements We would like to express our thanks to Dr. Paul Rayson for providing us access to Wmatrix's web interface and API, and Prof. Mike Thelwall for providing us with the SentiStrength Java version.

References

1. Abel F, Hauff C, Houben GJ, Stronkman R, Tao K (2012) Twitcident: fighting fire with information from social web streams. In: Proceedings of the 21st international conference companion on world wide web, WWW '12 companion. ACM, New York, pp 305–308
2. Achrekar H, Gandhe A, Lazarus R, Yu SH, Liu B (2011) Predicting flu trends using twitter data. In: Proceedings of the 2011 IEEE conference on computer communications workshops (INFOCOM WKSHPS), pp 702–707
3. Ahlqvist T, Halonen M, Heinonen S (2007) Weak signals in social media. Report on two workshop experiments in futures monitoring. SOMED foresight report, 1
4. Asur S, Huberman BA (2010) Predicting the future with social media. In: Proceedings of the 2010 IEEE/WIC/ACM international conference on web intelligence and intelligent agent technology, WI-IAT '10. IEEE Computer Society, Washington, DC, pp 492–499
5. Baron A, Rayson P (2008) Vard2: a tool for dealing with spelling variation in historical corpora. In: Proceedings of the postgraduate conference in Corpus linguistics
6. Bodendorf F, Kaiser C (2009) Detecting opinion leaders and trends in online social networks. In: Proceedings of the 2nd ACM workshop on social web search and mining, SWSM '09. ACM, New York, pp 65–68
7. Bollen J, Mao H, Zeng X (2011) Twitter mood predicts the stock market. J Comput Sci 2(1):1–8
8. Castells M (2012) Networks of outrage and hope: social movements in the Internet age. Polity Press/Wiley, Malden/Hoboken
9. Chang CC, Lin CJ (2011) Libsvm: a library for support vector machines. ACM Trans Intell Syst Technol 2(3):27:1–27:27

10. Charitonidis C, Rashid A, Taylor PJ (2015) Weak signals as predictors of real-world phenomena in social media. In: Proceedings of the 2015 IEEE/ACM international conference on advances in social networks analysis and mining 2015, ASONAM '15. ACM, New York, pp 864–871

11. Conway M, Doan S, Kawazoe A, Collier N (2009) Classifying disease outbreak reports using n-grams and semantic features. Int J Med Inform 78(12):e47–e58. Mining of Clinical and Biomedical Text and Data Special Issue

12. Diani M, McAdam D (2003) Social movements and networks: relational approaches to collective action. Comparative politics series. Oxford University Press, Oxford

13. Forman G (2003) An extensive empirical study of feature selection metrics for text classification. J Mach Learn Res 3:1289–1305

14. Forsyth DR (2009) Group dynamics. Cengage Learning, Wadsworth

15. Garside R, Smith N (1997) A hybrid grammatical tagger: Claws4. Corpus annotation: linguistic information from computer text corpora, pp 102–121

16. Gonzalez-Bailon S, Borge-Holthoefer J, Rivero A, Moreno Y (2011) The dynamics of protest recruitment through an online network. Sci Rep 1. http://dx.doi.org/10.1038/srep00197

17. Granovetter MS (1973) The strength of weak ties. Am J Sociol 78(6):1360–1380

18. Hall M, Frank E, Holmes G, Pfahringer B, Reutemann P, Witten IH (2009) The WEKA data mining software: an update. SIGKDD Explor Newsl 11(1):10–18

19. Haythornthwaite C (1996) Social network analysis: an approach and technique for the study of information exchange. Libr Inf Sci Res 18(4):323–342

20. Li R, Lei KH, Khadiwala R, Chang KCC (2012) Tedas: a twitter-based event detection and analysis system. In: Proceedings of the 2012 IEEE 28th international conference on data engineering (ICDE), pp 1273–1276

21. Lim SL, Quercia D, Finkelstein A (2010) Stakenet: using social networks to analyse the stakeholders of large-scale software projects. In: Proceedings of the 32nd ACM/IEEE international conference on software engineering, ICSE '10, vol 1. ACM, New York, pp 295–304

22. Mathioudakis M, Koudas N (2010) Twittermonitor: trend detection over the Twitter stream. In: Proceedings of the 2010 ACM SIGMOD international conference on management of data, SIGMOD '10. ACM, New York, pp 1155–1158

23. Piao SS, Rayson P, Archer D, McEnery T (2005) Comparing and combining a semantic tagger and a statistical tool for MWE extraction. Comput Speech Lang 19(4):378–397. Special issue on multiword expression

24. Prentice S, Taylor PJ, Rayson P, Hoskins A, O'Loughlin B (2011) Analyzing the semantic content and persuasive composition of extremist media: a case study of texts produced during the Gaza conflict. Inform Syst Front 13(1):61–73

25. Prentice S, Rayson P, Taylor PJ (2012) The language of islamic extremism: towards an automated identification of beliefs, motivations and justifications. Int J Corpus Linguis 17(2):259–286

26. Rad AA, Benyoucef M (2011) Towards detecting influential users in social networks. In: International conference on E-technologies. Springer, Berlin/Heidelberg, pp 227–240

27. Rashid A, Baron A, Rayson P, May-Chahal C, Greenwood P, Walkerdine J (2013) Who am i? Analyzing digital personas in cybercrime investigations. Computer 46(4):54–61

28. Rayson P (2008) From key words to key semantic domains. Int J Corpus Linguis 13(4):519–549

29. Rayson P, Garside R (2000) Comparing corpora using frequency profiling. In: Proceedings of the workshop on comparing corpora, WCC '00. Association for Computational Linguistics, Stroudsburg, pp 1–6

30. Rayson P, Archer D, Piao S, McEnery A (2004) The UCREL semantic analysis system. In: Proceedings of the beyond named entity recognition semantic labelling for NLP tasks workshop, pp 7–12

31. Sakaki T, Okazaki M, Matsuo Y (2010) Earthquake shakes twitter users: real-time event detection by social sensors. In: Proceedings of the 19th international conference on world wide web, WWW '10. ACM, New York, pp 851–860

32. Sankaranarayanan J, Samet H, Teitler BE, Lieberman MD, Sperling J (2009) Twitterstand: news in tweets. In: Proceedings of the 17th ACM SIGSPATIAL international conference on advances in geographic information systems, GIS '09. ACM, New York, pp 42–51
33. Taylor PJ, Dando CJ, Ormerod TC, Ball LJ, Jenkins MC, Sandham A, Menacere T (2013) Detecting insider threats to organizations through language change. Law Human Behav 37(4):267–275
34. Thelwall M, Buckley K, Paltoglou G, Cai D, Kappas, A (2010) Sentiment strength detection in short informal text. J Am Soc Inform Sci Technol 61(12):2544–2558
35. Thelwall M, Buckley K, Paltoglou G (2011) Sentiment in twitter events. J Am Soc Inform Sci Technol 62(2):406–418
36. Witten IH, Frank E, Hall MA (2011) Data mining: practical machine learning tools and techniques, 3rd edn. Morgan Kaufmann Publishers Inc., San Francisco
37. Yang Y, Pedersen JO (1997) A comparative study on feature selection in text categorization. In: Proceedings of the fourteenth international conference on machine learning, ICML '97. Morgan Kaufmann Publishers Inc., San Francisco, pp 412–420
38. Yu B (2008) An evaluation of text classification methods for literary study. Literary Linguis Comput 23(3):327–343

Chapter 8
Discovery of Structural and Temporal Patterns in MOOC Discussion Forums

Tobias Hecking, Andreas Harrer, and H. Ulrich Hoppe

1 Introduction

Discussion forums are a common element in massive open online courses (MOOCs). Since online courses with a large amount of participants cannot be supported individually by a tutor, discussion forums are often used for information seeking and information giving among the participants themselves. In previous studies it was shown that only a small fraction of all course participants actively take part in forum discussions. However, those who frequently participate in activities and possibly complete the course, which is also only a small fraction of all participants [6], are much more likely to be active in the forum [2]. Recent studies also relate the forum activity of course participants to influence in the community of learners [35]. This leads to the assumption that discussion forums are a communication channel for knowledge exchange between the core cluster of participants who are really willing to finish the course. This and the fact that course forums are more restricted in the topics that can be discussed makes these types of discussion forums an interesting case to study phenomena of knowledge exchange in electronic communities.

The goal of this work is to uncover the structure of knowledge exchange in MOOC discussion forums adapting mixed methods. The work is a significant extension of our previous work [17], incorporating new datasets and refining methodology. There exist several studies on forums for knowledge in different research areas including network analysis. However, most of the existing approaches

T. Hecking (✉) • H.U. Hoppe
University of Duisburg-Essen, Duisburg, Germany
e-mail: hecking@collide.info; hoppe@collide.info

A. Harrer
University of Applied Sciences Dortmund, Dortmund, Germany
e-mail: andreas.harrer@fh-dortmund.de

© Springer International Publishing AG 2017
J. Kawash et al. (eds.), *Prediction and Inference from Social Networks and Social Media*, Lecture Notes in Social Networks, DOI 10.1007/978-3-319-51049-1_8

analyse a static snapshot of a forum. This paper extends this body of research by investigating individual behaviour as well as interpersonal relationships between actors with a special focus on temporal dynamics, i.e. changes of behaviour and roles of actors over time. It aims at exploring the full process of structural analysis of discussion forums including network extraction, characterisation of individuals, and uncovering the latent meso-level structure of the communication networks. Thus, our approach covers a range of analytical methods combining text analysis and social network analysis techniques. This process includes the following three steps:

(Step 1) The first crucial step in applying network analysis methods to forum data is the extraction and modelling of a social network from forum threads that reflects directed knowledge exchange relations between actors. This is not a trivial task since the fact that two actors are active in the same discussion thread does not imply that these two also share knowledge in the sense that one of both replies to an information request of the other. Thus, the first step is to identify information seeking posts and related information giving posts in the discussion threads of the forums based on textual and structural features. This leads to a directed network of forum posts where each post that provides information points to one or more information seeking posts. As a second step, this network is transformed into a directed network of actors where the edges represent information giving relations. All edges carry timestamps, which allows for splitting the network into time slices that represent the information seeking/information giving structure of the forum communication in a particular period of time.

(Step 2) The extracted networks are analysed on the level of individuals in terms of trajectories of information seeking and information giving behaviour. Clustering actors according to similar trajectories yields interesting insights into the development of forum actors during the online course. Therefore, we define measures that properly characterise the information seeking and information giving behaviour of forum users taking into account, both the number of connections they have and their post quantity.

(Step 3) In discussion networks where potentially everyone can talk to everyone else there is no obvious structure or network topology. Especially in networks of the size of the ones under consideration and larger, it is almost impossible to make statements on the possible latent organisation of the network structure. For this reason different approaches can be applied to reduce the network to a macro-structure that captures the interaction pattern between components of the network, and thus, allow for a better interpretability of the latent organisation of the network. In particular, blockmodelling and tensor decomposition methods are modified and evaluated to perform this task for dynamic networks. Eventually, the discovered macro-structures are mapped to a knowledge exchange graph that depicts information flow between clusters of actors over time.

The methods are evaluated with and applied to anonymised datasets of two discussion forums of MOOCs offered on the Coursera platform[1] which are described in more detail in [29]. The first course is on "Introduction to Cooperate Finance" conducted during 11/2013 and 12/2013. Overall there were 8336 posts in 870 different threads by 1540 different actors. The second course is on "Global Warming: The Science and Modeling of Climate Change" with a discussion forum comprising 1007 actors with 5546 posts in 1020 different threads. For both datasets the forum activity peaked in the beginning and decreased afterwards until the end of the course which is typical for a MOOC discussion forum. In the following we refer to this dataset as "Corporate Finance" or "Global Warming", respectively.

The paper is structured as follows: After this introduction, Sect. 2 gives an overview of the related work of discussion forum analysis and the methodological background of this paper. The mentioned analysis steps are discussed separately in Sects. 3, 4, and 5. Results of the application of the developed methods to the two datasets are presented in Sect. 6. Section 7 concludes the paper and gives an outlook on possible directions of further research.

2 Background

2.1 Analysis of Discussion Forums

Research on discussion forums can be classified into content related and structure related analysis. Content related analysis deals with the content of the forum posts. Typical tasks are post classification, and discussion disentanglement. Especially in the case of MOOC discussion forums it is of huge interest to identify content related threads in which exchange of knowledge between participants takes place [8, 29] as well as the estimation of discussion quality [20].

In contrast to the content related analysis where the individual posts or threads are the main object of inquiry, structural analysis aims to model relations between entities in discussion forums in order to answer questions on knowledge diffusion through forum communication. One of the first studies on online mass communication by Whittaker et al. described dependencies between different properties of Usenet clusters such as thread depth, message length, and demographics [33].

Based on the information who gives information to whom in a question answer forum, one task is the identification of expert actors. Measures of expertise can be based on the quantity of questions and answers actors post to a forum or their position in the Q/A communication network [37]. Adamic et al. [1] draw upon this research by investigating forum data from Yahoo answers. They took a deeper look into the structural properties of networks between information givers and information seekers as well as their interest in topics. The gained knowledge could

---

[1] https://www.coursera.org/

be applied to distinguish different types of discussions and best answer prediction. Most recently, Gillani et al. [14] compared the structure of social networks extracted from different MOOC forums according to vulnerability and information diffusion. They could show that the cohesiveness of the networks depend on very few actors for most of the investigated forums.

An initial challenge that usually arises when network analysis methods are applied to forum data is the modelling of the underlying network. Especially in forums that have a non-nested thread structure, it is a challenge to establish relations between posts and consequently also between posters. While in forums like Yahoo answers, the relations between questions and answers are directly observable by the thread structure; this is usually not the case in MOOC forums. This problem is well known in language processing [21], however, there are only a few studies in network analysis research that tackle this problem explicitly, for example, [14, 28]. Thus, the study in this paper combines both the content related and structure related properties of discussion forums to model the underlying social network.

2.2 Identification of Roles in Communication Networks

Role modelling is of particular interest to characterise actors, for example, peripheral participants or "lurker" or active advice givers in the community [11]. The identification of the fundamental structures and topologies of networks is a means to understand the nature of interaction and actors' behaviour in complex networks. Role models cluster actors based on their position and connection patterns in the network. Thus, a cluster can be interpreted as users with similar role in the network. In contrast to community detection actors in the same cluster do not necessarily have to be densely connected within their cluster but only sparsely to outsiders. Moreover, role models do not require any connections between actors of the same cluster at all, although they are not forbidden. The goal is to identify connection patterns between clusters, which reduces the complex network structure to an interpretable macro-structure that reflects latent connection patterns between different roles of actors in the network.

2.2.1 Blockmodelling

Analyses that group with similar actors are subsumed as "positional analysis" [31]. One of the prominent approaches in that strand is "blockmodeling" [9] that assigns actors to clusters of similar relations with other actors and reduces the actor network to a more coarse grained network of clusters that represents the original network structure on a higher level (represented by an image matrix or block matrix). Those patterns can also be detected when analysing the algebraic composition of multiple relations [27], which is especially well suited for understanding the fundamental structure of a network when combined with a positional network that already reflects the basic structure of the network.

Fig. 8.1 Different types of blocks (relations between clusters)

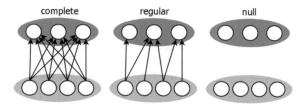

Connection patterns and role structures can be derived from different notions of equivalence and similarity between nodes [23]. Two important notions of equivalence of actors in a social network are structural and regular equivalence [31]. Two actors are structural equivalent if they have exactly the same neighbours in the network and consequently are not separated by more than two steps. A partitioning of the actors of a network based on structural equivalence results in a blockmodel with only complete or null relations between clusters (Fig. 8.1). This definition is usually too strict for typical sparsely connected information exchange networks. Thus, a more appropriate notion of equivalence is regular equivalence. Two actors are regular equivalent if they have the same connections from and to equivalent actors [32]. If there exists a regular relation from a cluster c_i to a cluster c_j, all actors in c_i point to at least one actor in c_j and all actors in c_j have at least one ingoing relation from actors in c_i. Consequently, regular equivalence blockmodels only contain regular or null relations (Fig. 8.1). Note that complete relations are a special case of regular relations. The different types of relations between clusters can be expressed by a $|C| \times |C|$ image matrix, where $|C|$ is the number of clusters. Each element of the image matrix gives the type of relation found between each pair of clusters.

Using a notion of equivalence to partition the network is often too strict and can lead to unrealistic results. Finding a partition of a network into clusters of regular equivalent nodes likely can result into an inappropriate high number of small clusters comprising only one or two nodes or only one cluster comprising all nodes (if the network does not contain sources and sinks). Thus, different relaxations can be applied [34]. On the one hand, one can relax the assignment of nodes to clusters such that each node has assigned a weight for each cluster that reflects to what extent the node belongs to this cluster [34]. A more common way is to relax the equivalence requirement for the nodes in the same cluster to similarities such that nodes within the same cluster should be almost structural or regular equivalent [9]. This is also the line followed in this paper. To measure the goodness-of-fit of such approximate blockmodels one has to compare the derived block structure with an ideal model that perfectly matches the desired relations, e.g. regular equivalence. The error of such approximate blockmodels is measured according to the minimum number of modifications (adding and deletion of edges) to be made such that the blockmodel is ideal.

Finding a nontrivial partition of nodes into k clusters such that they resemble an ideal blockmodel as good as possible is an NP-hard problem [34]. Different approaches exist to approximate an optimal solution [9]. The direct approach aims to optimise an initial partition by iteratively moving nodes from one partition to another or switching the clusters of two nodes such that the blockmodel error is continuously minimised. This procedure is computationally expensive and not feasible for large networks. Alternatively, one can compute the extent of regular equivalence of nodes (regular similarity) beforehand and cluster the nodes into clusters using arbitrary clustering algorithms. This indirect approach is much faster than direct optimisation but does not guarantee to find a local minimum of the blockmodel error.

2.2.2 Tensor Decomposition for Role Modelling

Apart from the complexity of the direct optimisation of a block structure, there are other drawbacks of blockmodels. First, traditional blockmodelling approaches tend to focus very strongly on the typical core-periphery structure of forum networks comprising a large cluster of active actors and many small clusters of sparsely connected actors. This is, on the one hand, a reasonable macro-structure but might not reflect the possibly latent interactions between different clusters of actors. Alternative approaches for mapping the network structure to a higher-order structure are based on tensor decomposition methods. These approaches are well suited for dynamic networks since the adjacency matrices of successive time slices of the network can be stacked to a third-order tensor (see left side of Fig. 8.2). These approaches are not only successfully used in relational learning tasks such as link prediction [10] and community mining [12] but are also applicable for role modelling [22, 25]. In contrast to blockmodelling, these methods do not optimise the assignment of nodes to clusters/roles towards fitting a target tensor that reflects an ideal block structure. Moreover, these methods optimise the partition towards the tensor representation of the evolving network itself.

Two related methods that are suitable for modelling dynamic and asymmetric relations between latent clusters are RESCAL [25] and DEDICOM [3]. The adjacency matrices for each time slice of an evolving network are modelled as a third-order tensor $X \in \mathbb{R}^{(|Act| \times |Act| \times T)}$, where Act is the set of actors in the network and T the number of time slices. The identification of actor roles and relations between them is performed simultaneously by finding a good fitting decomposition of X such that the following Eqs. (8.1) and (8.2) are minimised:

$$min_{A,R,D} \sum_{k=1}^{T} ||X_k - A \times D_k \times R \times D_k \times A^T||_F \qquad (8.1)$$

for DEDICOM and for RESCAL:

$$min_{A,R} \sum_{k=1}^{T} ||X_k - A \times R_k \times A^T||_F \qquad (8.2)$$

Fig. 8.2 Graphical depiction
of the RESCAL
decomposition

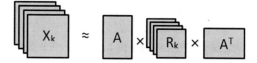

RESCAL can be considered as a relaxation of the DEDICOM model since it allows for varying relations among the latent roles over time. Since it is expected that relations between roles in discussion forums are also dynamic, in the following, this work is restricted to the RESCAL decomposition of the adjacency tensor. Graphically this decomposition can be depicted as in Fig. 8.2. The decomposition of the tensor can be efficiently computed using algorithms based on alternating least squares that update the matrices A, R_k, D_k in alternating fashion by minimising the objective functions in Eqs. (8.1) and (8.2). For more details of the concrete procedure we refer to [3, 25].

The rows of the matrix $A \in \mathbb{R}^{|Act| \times |C|}$ correspond to the actors in the network and the columns to the clusters. Each element of $a_{i,j}$ of A indicates a loading for actor i for a latent factor (cluster) j. The matrices $R_k \in \mathbb{R}^{|C| \times |C|}$ can be interpreted as the latent relations between the clusters $c_j \in C$ for each time slice. Consequently, the matrices can be seen as a specific notion of an image matrix as used in blockmodelling. If two nodes have similar loadings for the latent factors according to the matrix A, they have similar relations to other clusters or roles over time according to the relation tensor R. This bears some similarity with blockmodelling based on regular similarity in dynamic networks. The combination of both will be described in more detail in Sect. 5.

2.2.3 Estimating the Number of Clusters

All the investigated methods require a pre-specified number of clusters (roles). There is no general rule for good choices of this parameter and it has even been stated that this decision is "more an art than a science" [3]. The blockmodel error tends to decrease with growing number of clusters. The reason is that the more and smaller the clusters are the easier it is to establish nearly regular relations with a small number of relations between two clusters. If in an extreme case the number of clusters is equal to the number of nodes (each cluster contains one node), the blockmodel error is 0 since the set of nodes induce a perfect (but not desirable) regular equivalence partitioning. Moreover, the parameter specification should be reasonable for the goal of the analysis. Since the identified macro-structure models should be interpretable, typically one uses domain knowledge and hypotheses to decide on the number of clusters. This leads to a pre-specification of the structure (types of relations, number of roles) one aims to identify. This can also be extended to a deductive model fitting approach [9, pp. 233–244].

3   Network Extraction from Forum Posts

Starting with the lists of posts for each discussion thread, the goal of the network extraction step is to model a directed knowledge exchange network between the forum actors that reflects the information giving relations between them as good as possible. Each post entry contains information of the discussion thread it belongs to, the sub-forum, the (anonymised) actor, and the post content. There are two types of posts, namely regular posts and comments. Regular posts occur in linear sequence without sub-threading. In Coursera forums, actors can comment on regular posts but it is not possible to comment on a comment. The result is a flat thread structure in which regular posts are arranged in linear fashion, and comments always have a single parent post. While it is easy to relate a comment to its parent post, it is more complicated to relate a regular post to previous regular posts. In the following the network extraction steps are described in detail.

3.1   Forum Post Classification

Since not every forum post can be related to knowledge exchange, in a first step, information seeking and information giving forum posts have to be identified. There exist several tag sets for classes of forum posts which are often very fine grained, e.g. differentiating between initial questions and repeated questions and different types of answers [21]. However, since we are only interested in persons who have provided information or asked for some information in the forum, no further distinction of different types of information seeking and information giving posts is made. The result is a simplified tag set comprising "information seeking", "information giving", "social posts", and "others". Information seeking posts are course content related questions and problem descriptions. Information giving posts subsume all posts that provide some content related information. Apart from that, the forum is also extensively used by people who search for study clusters and general discussions. Those posts are labelled as "social" posts. All other posts that do not fit in any of the other categories are classified as "others". This classification also goes along with the observations made in [30]. Information seeking and information giving posts are the only relevant categories for the following studies. Social posts and others are only used for the purpose of filtering. Since discussion forums in Coursera courses are organised in sub-forums, it is easy to filter "social posts" since they usually occur in dedicated sub-forums that are not used for information exchange. The content related discussions take place in sub-forums that especially target content related issues regarding lectures and assignments. Thus, automatic classifiers are trained only to detect "information seeking", "information giving", and "others" posts from these types of sub-forums. Each forum post is represented by a vector of features adapted from [21]. In addition more content related features and a flag that indicates whether a post is a regular

Table 8.1 Feature set for post classification

Feature	Description
Forum ID	Not used for training but to restrict the automatic classification to sub-forums dedicated for course content related exchange.
Lexical similarity with initial post	Initial posts are often questions. High cosine similarity of the word vectors of a post and the initial post after stopword removal is an indicator for an information giving post.
Votes	Number of votes for each post. In Coursera actors can rate posts of other actors. Usually only content related posts receive votes whereas "other" posts like "Thank you for your answer" are unlikely to receive many positive votes.
Order	Position in the thread. The first post has position 0. The second 1. The position of a comment is its position in the commend chain + the position of its parent post. Information seeking posts often appear at the beginning of a discussion thread followed by information giving posts.
Is comment	Is the post a regular post or a comment on a previous post?
Length	Number of words of the post after stopword removal.
Question mark	Question marks are a good indicator for information seeking posts.
Exclamation mark	Exclamation marks occur often in information seeking and information giving posts, but not in "other" posts.
FiveWoneH	Number of occurrences of "Why", "What", "Where", "When", and "How". A high number indicates information seeking posts.
Special phrases	Phrases that indicate specific post types, i.e. variations of: help_me (e.g. "need help"), help_you, thank, did not, similar (e.g." same problem"), wrong.

post or a comment were added. Content related features are different variations of indicator phrases that suggest different post types. While phrases mapped to the "help_me" indicator phrase such as "need help" clearly indicate an information seeking post, other phrases such as "thanks" indicate a post that can be labelled as "others" such as "Thank you for your help". Table 8.1 gives an overview of the features used to encode each forum post. This list results from a larger list of structural and content features. A good selection and combination of features was derived by applying a genetic algorithm [24]. Starting with 30 random combinations tenfold cross validation was used as fitness function to assess the quality of a combination. A new generation of subsets of features are created out of the best 25% of previous combinations performing tournament selection [24]. The result is the best combination found in 30 generations. (Higher number of generations does not lead to further improvements)

A bagged random forest classifier [5] yielded the best results. The classification model was trained on 500 posts that were hand-classified by three experts by taking the majority of manual assigned classes for each post (interrater agreement according to Fleiss-Kappa $\kappa = 0.78, p < 0.005$). The classification has been evaluated using tenfold cross validation. The F1-score for the classification of

Table 8.2 Precision and recall of the post classification based on tenfold cross validation

	Precision	Recall
Information seeking	0.81	0.74
Information giving	0.61	0.73
Other	0.68	0.53

information seeking posts is good (F1-score = 0.77) and acceptable for information giving posts (F1-score = 0.66). However, posts of type "other" often lead to misclassifications as Table 8.2 shows. For this reason, the final classification was done by an iterative classification procedure (c.f. [26]). This algorithm uses an additional classifier trained on a dataset where the types of preceding posts of each posts are known. This makes the classification of information giving posts easier since those posts must have at least one preceding information seeking post. An initial classification is retrieved as described before. Then the additional classifier is applied to the data with the initially assigned class labels. This increases F1-scores for information giving posts increases to 0.71 and for information seeking posts to 0.79 based on evaluation on another set of 200 hand-classified posts.

3.2 Network Extraction

After post classification, the next task is to extract the knowledge exchange network. First, all posts that are not classified as either information giving or information seeking are removed from each thread leaving only question answering threads. The discussion threads in Coursera forums do not have nested sub-threads (except comments on posts), and thus, it is not directly visible from the thread structure which post refers to which previous posts. However, it is possible to attach comments to particular posts. Comments are often used to refer to older posts in the thread when the discussion leading to some kind of sub-thread. Thus, a network of posts is built according to the rules described in the following:

- **Information seeking/giving sequence:** A sequence of information seeking post is usually succeeded by a sequence of information giving posts. After the information giving sequence sometimes a new information seeking sequence starts followed by another information giving sequence. Thus, a thread can be decomposed into a set of such information seeking/giving sequences. The first rule applied is to link each of the information giving post to the information seeking posts in each information seeking/giving sequence in a thread.
- **Comments as sub-threads:** Comments are attached to a parent post. Comments attached to a parent post are treated as regular posts of a sub-thread with the parent post as initial post. Then the sequence rule described before is applied to this sub-thread.

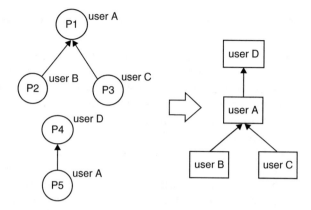

Fig. 8.3 Mapping a post network to a network of actors

The comparison of the extracted relations by the rules and the relations identified by the human classifiers yields high accuracy (F1-score = 0.89). The performance achieved by the rules described above indicates that actors in the investigated MOOC forum usually maintain the structure of a thread themselves, using comments instead of regular posts if they refer to much earlier posts. Consequently the discussion structure in threads is usually not very entangled and the simple rules are very effective for network extraction. We also tried to incorporate the lexical overlap between the posts as it was done for the case of chat disentanglement [18], but this did not improve the results. A reason can be that discussion threads structured according to discussion topics already induce a lexical overlap of many of the contained posts such that these features do not provide new information.

In the post network resulting from the previous step each post node is annotated with the author of the post and a timestamp. Furthermore, the network is highly disconnected since links are based on individual forum threads. Next, each post node labelled with the same author is collapsed into a single node representing the actor resulting in the final knowledge exchange network between forum actors, similar to [16]. Figure 8.3 gives an example.

The actors who post neither information giving nor information seeking posts do not appear in the extracted network since "social" and "other" posts were ignored. After deletion of isolated actors and the actor representing all anonymous posts, the resulting network for the "Cooperate Finance" course comprises of 647 actors who actively participated in content related forum discussions by asking for information or providing information to information seekers. The network is very sparse with only 1303 edges. For the smaller course, "Global Warming", there remain 348 actors as nodes in the extracted network. The actors in this network are more densely connected (1291 edges) than in the previous one.

4 Individual Development: Behavioural Roles over Time

In the following we present the analysis approach for identifying patterns of individual development of forum actors over time. The goal is to characterise actors according to their information seeking and information giving behaviour. The question is whether there exist actors who are important in the sense that they either provide much information (information giving) or ask important questions or raise problems that stimulate the community to respond (information seeking). Simply counting the number of information giving and information seeking posts of each actor can be too simple. Someone who has a high number of information giving posts might not necessarily be a real information giver in the community since the high number of posts can result from a long discussion with one single actor [37]. On the other hand, taking the out-degree or in-degree centrality as a characteristic measure can also lead to unreliable results since a high degree can result from one single post. Consequently, an information giver should have many information giving posts reaching many different information seekers and an information seeker should have many information seeking posts that are replied by many other actors. Thus, our solution is to use a combined measure of the number of posts and the diversity of connections resulting in two measures we refer to as outreach and inreach that will be defined below.

4.1 Definition of Inreach and Outreach

Given a single node of a weighted and directed network, the diversity of its in and outgoing relations can be characterised by a measure of entropy. Equations (8.3) and (8.4) calculate the diversity of outgoing and ingoing relations for a node i, where $w(e_{i,j})$ is the weight/multiplicity of an edge from i to j and $od(i)$ is the out-degree and $id(i)$ the in-degree of i (taking into account edge weights, which is equal to the number of its information giving posts or information seeking posts, respectively, except in rare cases where one information giving post matches multiple information seeking posts).

$$H_{out}(i) = \frac{-1}{od(i)} \sum_{j \in outneigh(i)} w(e_{i,j}) * log\left(\frac{w(e_{i,j})}{od(i)}\right) \qquad (8.3)$$

$$H_{in}(i) = \frac{-1}{id(i)} \sum_{j \in inneigh(i)} w(e_{j,i}) * log\left(\frac{w(e_{j,i})}{id(i)}\right) \qquad (8.4)$$

The value for the connection entropy of node i reaches its maximum, if all posts of i address different nodes and its minimum 0 on the other extreme. In order to combine diversity and posting activity the number of corresponding posts of node i

($= od(i)$ for information giving, $= id(i)$ for information seeking) can be multiplied with $(H_{out}(i) + 1)$ or $(H_{in}(i) + 1)$, respectively, resulting in Eqs. (8.5) and (8.6) for the outreach and the inreach.

$$outreach(i) = od(i) - \sum_{j \in outneigh(i)} w(e_{i,j}) * log(\frac{w(e_{i,j})}{od(i)}) \qquad (8.5)$$

$$inreach(i) = id(i) - \sum_{j \in inneigh(i)} w(e_{j,i}) * log(\frac{w(e_{j,i})}{id(i)}) \qquad (8.6)$$

As a result, the outreach of actor i is at minimum the number of its information giving posts, if all posts address the same actor. The statement is similar for the inreach of an actor but with respect to its ingoing (information receiving) relations. Consequently the two described measures are helpful to detect active actors that reach or are reached by many other actors in the network. The comparison of the outreach and inreach of actors allows for a characterisation of the actors with respect to their behaviour, i.e. information seeking and information giving.

4.2   In- and Outreach over Time: Identification of Characteristic Actor Trajectories

The behaviour of an actor in a course forum usually will change over time. To cope with this, the next step after the definition of the in- and outreach measures is to uncover typical trajectories of information giving and information seeking behaviour of actors over time. Since the edges of the evolving knowledge exchange network carry timestamps it is possible to calculate the in- and outreach of the actors at different times by splitting the dynamic network into successive time slices. However, calculating these measures for each time slice independently does not account for a possible long term effect of actor activities. Consequently the calculation of an in- and outreach trajectory of an actor has to take into account the history of the actors behaviour as well. In time slice t of the network older edges should not be weighted as high as recent edges but can still have an effect. In order to achieve this we apply a growing window approach to model the evolution of the knowledge exchange network at different points in time. The approach includes a linear forgetting function which weights the edges according to their recentness as suggested in [7]. The parameter Θ can be used to control the extent of decline of edge weight. Other weighting functions, for example, exponential decline of edge weights are also possible. The weighted adjacency matrix $wAdj_t$ of the tth time slice is then calculated as a weighted sum of the unweighted adjacency matrices Adj_i of the network in the time slices up to t as in Eq. (8.7).

$$wAdj_t = \sum_{i=1}^{t} Adj_i * \Theta^{t-i}, \Theta \in (0, 1] \tag{8.7}$$

The calculation of in- and outreach on each of the resulting weighted networks derived by this growing window approach results in an inreach sequence and outreach sequence for each actor. The growing window approach for modelling evolving networks is an additive procedure and the information of earlier time slices is not completely lost in later time slices. An actor who stops communicating in the forum has still a value greater than 0 for in- or outreach lowered by the forgetting function in each slice. Thus, the slices should be kept rather short to capture the dynamics of the changes of individual in- and outreach. In this work this size is fixed to 3 days since it can be assumed that a forum post receives most of the reactions within this time window [35]. The goal is to discover characteristic patterns in the actor behaviour over time based on the similarity of those sequences. A naive approach would consider the sequences as numerical feature vectors and apply, e.g. k-means clustering to reduce the set of sequences to a small number of clusters of similar sequences. However, using this approach, both the inreach and outreach sequences have to be considered simultaneously resulting in a high number of dimensions of the feature vector which can be problematic for traditional clustering methods. Thus, k-medoids clustering [19] is applied in an alternating fashion. First, clusters are derived by partitioning the actors according to their inreach sequences. In a second step the medoids of the found clusters are used to initialise the k-medoids clustering according to the outreach sequences. This procedure alternates until the clusters have stabilised.

The medoids of each cluster can be considered as its prototypical representative. Thus, this method is appropriate to reduce the vast amount of individual sequences to uncover an interpretable set of typical trajectories actors can have in the knowledge exchange forum. Results in Sect. 6.1 will show the utility of this approach to identify the emergence of different behavioural roles and their changes over time.

5   Macro-Structure of Evolving Knowledge Exchange Networks

In both of the dynamic knowledge exchange network communication links repeat very rarely over several time slices. More than 80% of the communication links between two actors occur only once. This leads to the assumption that in the forum there exist no stable cohesive subcommunities of actors over time. However, there can be actors who behave similar over time with regarding their connection patterns to others without necessarily having a direct connection. Thus, reducing the network to a macro-structure that reflects the information flow between different clusters of actors according to connection patterns can lead to interesting insights into the overall structure of the knowledge exchange in discussion forums. As described

in Sect. 2.2, for the discovery of those macro-structures in evolving networks
there exist different approaches, blockmodelling and tensor decomposition. Both
have certain advantages and disadvantages for the task. In the following the
two approaches are compared regarding the utility for the analysis of knowledge
exchange networks. We will also propose adaptations of existing methods that better
fit the needs for uncovering macro-structures of knowledge exchange.

In particular we aim to find regular relations between clusters. On the one hand,
complete relations based on structural equivalence, as described in Sect. 2.2, are too
strict for our sparse forum discussion network. On the other hand, regular relations
reflect information flow between clusters. For information giving relations between
actors, regular relations between clusters can be interpreted as existing information
flow from cluster c_i to cluster c_j. This does not require that everyone in cluster c_i has
to talk to everyone in cluster c_j.

5.1 Dynamic Blockmodelling

The task of finding a well-fitting block structure of a network as described in
Sect. 2.2 becomes even more complex in dynamic networks. An ideal blockmodel
for dynamic networks defines a partitioning of the nodes into k clusters such that the
nodes within the same cluster are almost regular equivalent in each time slice. For a
given dynamic network sampled in T time slices, the goal is then to find a partition
of the nodes that is as close as possible to an ideal dynamic blockmodel in each time
slice.

A simple approach for indirect blockmodelling in dynamic networks would be
to calculate the extent of regular equivalence (or regular similarity) $REGESim_t(i,j)$
for a pair of nodes i and j in each time slice t and take the average as in Eq. (8.8).

$$dynamicREGESim_{i,j} = \sum_{t=1}^{T} \frac{REGESim_t(i,j)}{T} \qquad (8.8)$$

For computing the regular similarity of all node pairs the REGE algorithm
[4] is applied. The resulting overall similarity can then be used as input for a
clustering algorithm that assigns the nodes to clusters. Here hierarchical clustering
is used. In the following this approach is referred to as SIDBM (Sequential Indirect
Dynamic Blockmodelling). However, focusing on each time slice separately has the
disadvantage that if the similarity of node pairs varies heavily over time, a high
similarity between two nodes in only one time slice can have a huge effect on the
outcome even if the nodes are very dissimilar in other time slices.

To reduce this problem, the incremental blockmodelling method for multi-
relational networks proposed by Harrer and Schmidt [15] can be utilised. Optimising
partitions across time slices of evolving networks can be considered as a special case

of multi-relational blockmodelling when the edges in each time slice are considered as edges of one particular relation. The approach identifies a blockmodel across multiple relations as follows:

1. Find a blockmodel for each relation (time slice) starting from a random partition.
2. Select the blockmodel that yields the smallest average error for each partition.
3. Find other blockmodels by optimising the partition found in step 2 for each time slice.
4. Repeat steps 2 and 3 for n iterations and select the blockmodel that yields the smallest error.

This approach guarantees to find a partition of the nodes according to a given equivalence relation with a local minimum of the average blockmodel error across time slices. However, the third step incorporates direct optimisation of a given network partition to fit a certain blockmodel which is computational expensive especially for many time slices, and thus, it is not feasible for large networks. We propose an adaptation of the method that keeps the original procedure of incremental updates of the node partition but uses the indirect (similarity based) clustering instead. In the first step the extent of regular equivalence the REGE algorithm [4] is applied to compute the extent of regular equivalence for each node pair in each time slice t. Instead of directly optimising the partitions in step 3, the similarity of the nodes in time slice $tdynamicREGESim_t(i, j)$ is computed as the average regular similarity of the nodes in the time slice t and the similarity of the nodes of the time slice s which yields the best fitting blockmodel across all time slices (see Eq. (8.9)).

$$dynamicREGESim_t(i, j) = \frac{REGESim_t(i, j) + dynamicREGESim_s(i, j)}{2} \quad (8.9)$$

In the next iteration again the similarity $dynamicREGESim_s$ that yields the best fitting blockmodel for all time slices is chosen to re-compute the similarities of node pairs according to Eq. (8.9). After a defined number of iterations (in this work 25) the best blockmodel will be returned. In the following, this method is referred to as IIDBM (Incremental Indirect Dynamic Blockmodelling).

5.2   Role Modelling Based on Tensor Decomposition

While the more traditional blockmodelling approaches optimise the partitioning of the nodes of an evolving network towards ideal image matrices for the time slices (or target tensor), RESCAL (see Sect. 2.2) aims to approximate the adjacency tensor itself by inferring the loadings of the nodes to latent clusters. Thus the results of link based clustering can differ much from an ideal regular equivalence blockmodel. In this aspect, the results are not as easy to interpret as traditional blockmodels. For example, the values of the relation matrices R_k (Eq. (8.2)) do not have clear semantics in terms of certain types of relations and node equivalences such as the typical image matrices produced by blockmodels. To overcome this problem we introduce a slight modification of the RESCAL approach for link based clustering that biases the clustering of the rows of the matrix A towards a given type of node equivalence. Instead of clustering the rows of the resulting matrix of loadings A directly, A is modified to A' by the following matrix multiplication (Eq. (8.10)):

$$A' = S \times A \tag{8.10}$$

$S \in \mathbb{R}^{|Act| \times |Act|}$ can be any similarity matrix between the nodes (actors) in the network. With this simple modification, the elements $a_{i,r}$ of the role matrix $A' \in \mathbb{R}^{|Act| \times |C|}$ contain high values if actor i has a high loading for cluster c_r and also a high similarity with other actors who also have high loading for cluster c_r. Consequently this approach biases the assignment of nodes to clusters towards a certain type of relations. Thus, in the following this adaptation of link clustering based on RESCAL is referred to as biased RESCAL and the matrix S is considered as the average regular similarity of node pairs as given by Eq. (8.8) in the section before. In order to allocate nodes to clusters uniquely any partitioning clustering (in this work hierarchical clustering) approach can be applied to the rows of the role matrix A'.

5.3   Formal Evaluation

Since there is no ground truth data for evaluation (i.e. no predefined classification of actors as in Bader et al. [3]), the described macro-structure models can only be verified by comparing the model to the actual data. One obvious measure is the blockmodel error that results by comparing the node partitioning to an ideal regular equivalence blockmodel comprising either regular or null blocks described in Sect. 2.2. Second, the discovered regular relations between clusters are evaluated with respect to the density of links between the related clusters.

As mentioned in Sect. 2.2 the number of clusters has to be decided based on empirical observations and hypotheses about the network structure. Based on previous studies it is reasonable to assume a core-periphery structure with a core of actors who are well connected and active in the forum [13, 35]. It can also be assumed that there are many peripheral nodes that appear either as information seeker or information giver only a very few times. Further, it is likely that there exist a (semi-peripheral) role comprising mostly information givers and a role comprising of mostly information seekers.

Thus, only results where the number of clusters was fixed to 5 are reported. Higher numbers scale the resulting blockmodel error but do not lead to changes in the ranking of the models. Further, the resulting models are more difficult to interpret in application scenarios. The size of a time slice was set to 2 weeks of forum communication resulting in 3 slices for the "Corporate Finance" forum and 4 slices for the "Global Warming" forum. This choice has been made since the most

Table 8.3 Error of regular equivalence blockmodels produces by different methods

	RESCAL	Biased RESCAL	SIDBM	IIDBM	Random partitioning
Corporate Finance	828.9	646.39	397.15	321.96	1027.71
Global Warming	872.71	437.01	504.56	447.02	1143.11

MOOCs are organised in thematic blocks of 1 or 2 weeks and it is likely that the forum communication is oriented towards this pace. It was shown that the size of time slices have a systematic effect on clustering outcomes and a good resolution can be achieved if the time slices are long enough to capture the typical duration of production cycles of the studied communities [36].

5.3.1 Fitting an Ideal Regular Block Structure

As already mentioned regular block structures are especially suited for the mapping of information exchange as evidence for existing information flow within one or between two clusters. Table 8.3 depicts the sum of blockmodel errors of the time slices with respect to an ideal regular equivalence blockmodel of the previously described algorithms.

It can be seen from Table 8.3 that the incremental indirect dynamic blockmodelling approach (IIDBM) yields better results than the simple sequential approach (SIDBM) for both datasets. For the "Cooperate Finance" discussion forum, this approach also gives the best fitting blockmodel. The tensor decomposition method RESCAL falls behind which is expectable since it does not explicitly optimise the portioning of the nodes according to regular equivalence. In contrast to that, the biased version of RESCAL leads to slightly better results than IIDBM for the "Global warming" dataset.

5.3.2 Density Patterns

Evaluating macro-structures according to the fit to an optimal regular equivalence does not allow for statements about density patterns between clusters of nodes since a regular relation between two clusters c_i and c_j is fulfilled if all nodes in c_i have at least one outgoing relation to one node in c_j and all nodes in c_j have one ingoing relation to one node in c_i. In the following the number of links between the cluster pairs for which a regular relation was discovered by the described macro-structure modelling methods is investigated in more detail. The density of links ρ_{c_i,c_j} between two clusters c_i and c_j is simply the fraction of links actually pointing from nodes in c_i to nodes in c_j, and the number of links that could exist between the two clusters.

Table 8.4 Average relation density ρ_C of different macro-structure models for the "Corporate Finance" discussion forum (number of relations classified as regular in brackets)

Time slice	RESCAL	Biased RESCAL	SIDBM	IIDBM
1	0.01 (17)	0.004 (13)	0.004 (9)	0.005 (6)
2	0.0003 (15)	0.002 (4)	0.002 (4)	0.0001 (8)
3	0.0003 (16)	0.0002 (8)	0.002 (16)	0.005 (4)

Table 8.5 Relation density ρ_C of different macro-structure models for the "Global Warming" discussion forum (number of relations classified as regular in brackets)

Time slice	RESCAL	Biased RESCAL	SIDBM	IIDBM
1	0.031 (15)	0.022 (9)	0.011 (10)	0.014 (8)
2	0.038 (17)	0.018 (9)	0.016 (8)	0.014 (9)
3	0.037 (12)	0.013 (7)	0.013 (6)	0.010 (8)
4	0.059 (15)	0.015 (9)	0.018 (6)	0.028 (6)

Let $\mathfrak{R}_{reg}$ be the set of ordered pairs of clusters $< c_i, c_j >$ for which a regular relation exists from cluster c_i to cluster c_j. The average relation density of the regular relations ρ_C for a given clustering C is then given by Eq. (8.11) where Adj denotes the adjacency matrix of the network.

$$\rho_C = \frac{1}{|\mathfrak{R}_{reg}|} \sum_{c_i, c_j \in C: <c_i, c_j> \in \mathfrak{R}_{reg}} \frac{1}{|c_i| * |c_j|} \sum_{l \in c_i, m \in c_j} Adj_{l,m} \qquad (8.11)$$

The average relation density is evaluated for all cluster pairs for which a relation is closer to be regular than to being non-existent. Results for the different macro-structure modelling methods are given in Table 8.4 for the time slices of the "Cooperate Finance" network and Table 8.5 for the "Global warming" network. The number of clusters was fixed to 5, as above. The numbers in brackets denote the number of role relations classified as regular.

It can be seen from Tables 8.4 and 8.5, that the average density of the relations induced by the RESCAL model is higher on average compared to the models that aim to optimise a regular equivalence blockmodel. Further, the number of relations that are considered approximately regular is much higher. The high blockmodel error for RESCAL reported in Table 8.3 indicates that the discovered regular relations are much more approximate than for other models. Overall the relations are denser for the "Global Warming" course which results from the higher network density (see Sect. 3.2). However, the values are overall very close to 0. An explanation is the sparsity of the networks extracted from the discussion forums when those networks are restricted to information giving relations.

5.3.3 Assessment of Methods

Previously it was shown that blockmodelling and tensor decomposition approaches can be used to uncover different types of macro-structure in knowledge exchange networks. Blockmodelling approaches that cluster nodes based on the extent of regular equivalence between pairs are a useful means to identify parts in the network with specific function. This type of modelling is more suited to identify functional parts of the network like core-periphery structures, clusters of experts or information seekers, etc. In contrast to that, the original RESCAL tensor decomposition model is more capable of reflecting the number of outgoing links that point from c_i into c_j while failing to partition the network according to regular relation and null relation patterns in the network. In general, it can be seen that blockmodelling methods perform much better on the task of fitting a regular relation structure between roles than tensor decomposition on the bigger and much more sparsely connected "Corporate Finance" network. On the other hand, in the smaller and more densely connected "Global Warming" network the biased version of RESCAL and IIDBM perform similar with respect to the blockmodel error and slightly better regarding density patterns. Especially in those networks with denser interrole relations the combination of RESCAL and regular similarity can be considered as a good alternative to the more traditional blockmodelling approaches since it combines, both, a good fitting blockmodel and density of relations. Consequently in the following IIDBM is used for the "Corporate Finance" and biased RESCAL for "Global Warming" dataset.

6 Applications

In the following, the utility of the described methods for modelling behavioural roles (Sect. 4) and structural roles (Sect. 5) are demonstrated by applying them to the networks extracted from the two online courses.

6.1 Trajectories of Behavioural Roles

Based on the k-medoids clustering of in- and outreach sequences of the actors described in Sect. 4.2 different behavioural roles for both of the investigated discussion forums can be identified. The medoids of the clusters are taken as prototypical examples for a set of actors with similar behaviour over time. The networks were sampled in short time slices of 3 days for the reasons explained in Sect. 4.2. The damping factor Θ for Eq. (8.7) for creating in- and outreach sequences was fixed to 0.9 leading to a slight linear decline of the values over time if an actor stops posting to the forum. A proper value for the number of clusters was estimated by optimising the separation of clusters according to the average silhouette width.

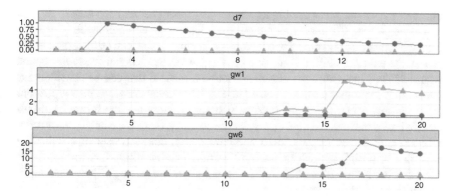

Fig. 8.4 Post and leave pattern (*top*). Late starter pattern (*bottom*) (*Red line*: inreach, *Cyan line*: outreach)

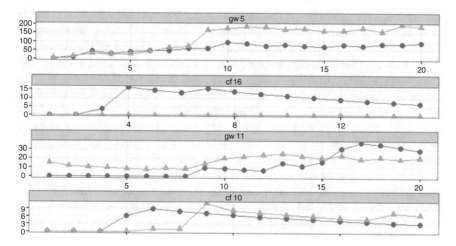

Fig. 8.5 Patterns of actors who are active during a longer period (*Red line*: inreach, *Cyan line*: outreach)

For the "Corporate Finance" forum 18 clusters and for the "Global Warming" forum 15 clusters were determined. Figures 8.4 and 8.5 depict the trajectories of some of the medoids representing different types of in- and outreach patterns.

The depicted trajectories reflect typical actor behaviour in the discussion forums of the online courses. A pattern that is frequent in both discussion forums is the "post and leave" pattern. 76% in the "Corporate Finance" MOOC and 78% in the "Global Warming" MOOC are clustered around medoids that represent this pattern. Representative trajectories are given by Fig. 8.4. Those actors post only once very little to the discussion forum and then become inactive. A special case of this behaviour can be considered as "late starting" as the example (gw6) from the "Global warming" forum shows. Those actors are only active in a very late phase of the course and have higher values for inreach or outreach than in the post and leave pattern. Clusters around those medoids are rather small in the "Corporate Finance" course so that only 7% of actors are in clusters that can be classified as late starter cluster. Late starting is much more frequent in the "Global Warming" course where 55% of the actors are clustered around a medoid that can be considered as late starter. Note that a cluster considered as late starter cluster is also a post and leave cluster. An explanation could be that some actors use the forum to discuss certain issues close to the final exam. On the contrary, there is a smaller set of actors who are active over longer periods and have a high outreach over time indicating expertise or a high inreach indicating that they use the discussion forum mainly for gathering information ("All time information giving/seeking). Typical examples drawn from the discovered cluster medoids are depicted in Fig. 8.5. The trajectory at the top of Fig. 8.5 (medoid of cluster gw 5) depicts an actor who is an all time information giver in the "Global Warming" forum. This actor has an extremely high outreach over time while also receiving lots of information from other actors during the course. This would be typical for course staff, but in this case, it is a regular actor who behaves like a tutor in the course constantly providing information to many others. On the contrary, the second example from the top of Fig. 8.5 (cluster cf 16) shows a typical cluster representative for the "Corporate Finance" course who has a much higher inreach than outreach over the entire period of observation. Those actors use the discussion forum mainly for gathering information. Very interesting patterns that are quite seldom in both datasets (7% in "Corporate Finance" and 17% in "Global Warming") are clusters comprising actors who develop from information seekers with higher inreach than outreach to information givers with higher outreach than inreach or vice versa as depicted by the two examples at the bottom of Fig. 8.5. The medoid of cluster gw 11 of the Global Warming course starts as someone who actively gives information to peers. In the last third of the course the actor develops into an active information seeker indicated by the increase of inreach. On the contrary, in the example at the bottom (cluster cf 10) the actor starts as an information seeker but develops into an information giver. The latter two cases are of particular interest since it reflects personal development through active forum participation.

In general the temporal pattern of individual development regarding clusters of in- and outreach sequences of the actors reflects the often reported observation that the majority of actors use the forum only occasionally and consequently do not produce many posts. On the other hand, there are some actors who are more active, and thus, often have a more diverse development over time.

6.2 Macro-Structures of Knowledge Exchange

In the following the actor roles and relations between them in the two discussion forums under investigation are modelled according to the methods described in Sect. 5 to derive blockmodels that map the macro-structures of knowledge exchange between clusters of actors that can be interpreted as roles. In Sect. 5.3 it was shown that it is difficult to find clusters of actors so that actors of one cluster have dense outgoing relations to actors of the same or another cluster. On the other hand, regular equivalence relations could be discovered to some extent. As stated before, those regular relations are considered as important for mapping knowledge flow between roles in the course forums. Thus, the blockmodels derived by the IIDBM method for the "Corporate Finance" and by the biased RESCAL method for the "Global Warming" course were used since these have the best fit for the two datasets in terms of a low blockmodel error and moderate relation density, respectively (see Sect. 5.3). Based on the considerations outlined in the beginning of Sect. 5.3 and the following evaluation the number of clusters roles was fixed to 5 and the networks were sampled in time slices of 2 weeks. This results in 3 time slices for the "Corporate Finance" and 4 time slices for the "Global Warming" dataset.

The knowledge exchange graphs of the particular time slices merged into a single representation are depicted in Fig. 8.6. Each cluster is considered twice per time slice, as sending and receiving cluster (depicted as pills). This allows for representing the structure of knowledge exchange between clusters across time slices (T1, T2, T3, T4) as a directed acyclic graph where the edges have a partial temporal order, which is beneficial for depicting information flow over time. An edge states that the relation between two clusters is almost a regular relation (imposed by regular similarity of the actors of the two clusters, respectively).

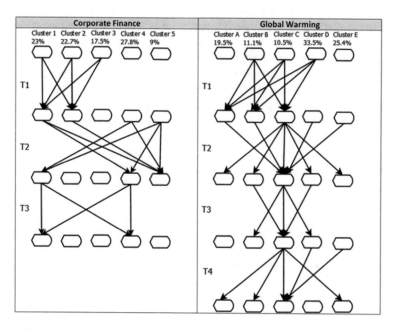

Fig. 8.6 Relations between clusters/roles over time. Percentage of actors is given for each cluster

The communication in both discussion forums is dominated by one core cluster (Cluster 1 (23% of all actors) in the "Cooperate Finance" forum and Cluster C (10.5% of all actors) in the "Global Warming" forum). These clusters not only comprise of actors who are active in each time slice and have many information giving relations to the other clusters but also receive information others. Interesting to see is that they have regular relations with themselves in most time slices. In time slices when peripheral clusters of actors have regular relations with other clusters they tend to connect with the core cluster. This is typical for core-periphery structures in social networks [9, pp. 235–236]. The concentration of the communication around a core is even more pronounced in the smaller and more densely connected "Global Warming" forum. There exists also a semi-periphery (cluster B and D). These clusters are active during the entire period but build connections mostly with the core cluster C. In this aspect, it can be assumed that in MOOC discussion forums information constantly circulates within a core cluster and occasionally alternates between this core cluster and the more peripheral clusters. Thus, the role of the core actors can be considered as especially important for the coherence of knowledge exchange.

Actors who are only connected to other clusters in particular time slices are primarily in clusters 3 and 5 in the "Corporate Finance" course and clusters A and E in the "Global Warming" course. This reflects the role of actors who show a "post and leave" behaviour and appear as information seekers or information givers only in particular time slices. A possible explanation is that those actors become active in the forum during course topics that are of interest or for which they encounter concrete problems. This actor role is much more salient in the "Cooperate Finance" course.

The knowledge exchange structures in both discussion forums further have in common that the most relations between roles exist in the first two time slices. This reflects the typical decrease in activity over time in MOOC discussion forums due to dropouts of the course and the "post and leave" behaviour reported in Sect. 6.1. This is especially the case in the "Corporate Finance" course. In the "Global Warming" forum, however, there exist relations from the core cluster to almost all other clusters in the last time slice. This is also reflected by the high fraction of "late starters" reported in Sect. 6.1 for this course. In the "Corporate Finance" course late starting can be associated with some actors in the cluster 4 which is the only active role in the last time slice except the core cluster 1.

In general, it can be said that the assignment of actors to clusters or different roles, respectively, by the IIDBM or biased RESCAL method mostly reflects the heterogeneous activity patterns of the actors. The flow graphs in Fig. 8.6 reveal that only a limited number of actors are available to establish (the desired) sustainable information exchange over time.

7   Conclusion and Further Work

In order to tackle the problem of identification and analysis of the structure of knowledge exchange in discussion forums one has to face several challenges. This paper presented a holistic approach incorporating network extraction and modelling from unstructured forum data, analysis of individual development, as well as an analysis of connection patterns in the forums of two Coursera MOOCs. A crucial step is the extraction of the underlying communication network from forum data taking into account the semantic of links. Although, the extraction of a directed network of actors who give information to each other using machine learning methods for forum post classification is not perfect, we expect that the resulting network reflects the true knowledge exchange structure much better compared to naïve approaches, for example, simply linking actors who appear in the same thread. In contrast to the most existing studies on network analysis on forum data time is explicitly taken into account enabling to investigate changes in the structure of the forum communication.

Based on the definition of the in- and outreach measures (Sect. 4), it is possible to characterise the posting behaviour of individuals, i.e. the role of actors in terms of information giving and information seeking. These measures combine the volume of posts with the diversification of connections in the extracted communication network. Analyses of the in- and outreach of actors over time in the presented case studies revealed different temporal patterns of posting behaviour. While the most actors are only partially active in the forum "post and leave", the more active users can be divided into persons who constantly act as information givers or information seekers or actors who change their behaviour over time. These later cases are of particular interest for the research on knowledge building in large scale online courses since these patterns are indicators for individual development.

On a meso-level, different methods for clustering actors based on the similarity of their connection patterns were developed (Sect. 5) to uncover meaningful structures from the extracted communication networks enhancing the interpretability of the knowledge exchange structure. The clusters found are not necessarily cohesive but reflect the global position of actors in the knowledge exchange network. Thus, they can be interpreted as roles. The connection patterns between these clusters change over time, which introduces further challenges to the role modelling. It was shown that the modification of multi-relational blockmodelling methods and role modelling based on the RESCAL tensor decomposition are useful to detect underlying structures in the knowledge exchange network of forum users. In both courses, knowledge exchange is typically dominated by a small number of actors who actively participate in the forum discussion during the course, which is in line with previous findings such as [14, 35]. Other actors are not active over the entire period but appear as information seekers or information givers in particular time slices where they establish connections with the core cluster. Nonetheless, these actors can be temporally important with respect to the coherence of the discussion since they might provide important information input into the forums and initiate longer discussions.

In general, the results of the analysis of individual development (Sect. 6.1) and the macro-structure of knowledge exchange (Sect. 6.2) yield similar conclusions but highlight different aspects of forum communication in the investigated discussion forums. While Sect. 6.1 focused on the individual behaviour of actors over time, the results of Sect. 6.2 provided insights into the overall structural characteristics of the communication network. We believe that such a combination of different methods can contribute to a better and more substantial understanding of the complex process of knowledge exchange in discussion forums of large scale online courses.

Similar patterns related to forum activity were identified for the networks of both courses even if they differ in size and density. This also supports conclusions made in related studies, for example [38], that the emergence of a sustainable and constructive knowledge exchange as well as individual development has to be much better supported. This can, for example, be achieved by providing better communication channels such as Q/A systems or by using techniques of social recommendation to support participants in finding proper communication partners. To improve those support mechanisms, in the future the nature of the knowledge exchange between clusters of actors with similar roles in the network should be further explored. It is also worth to investigate the impact of role relations as described in Sect. 6.2 more deeply on the content level. This could, for example, be done by incorporating the discussion topics in the analysis. This would yield further insights into content related communication in MOOC discussion forums and the diffusion of information within the community of course participants. Furthermore, the results can be combined with analyses of other aspects of online courses. For example, it would be interesting to see whether a participant who develops from an information seeker to an information giver or vice versa in the discussion forum also changes behaviour regarding other course activities, or whether there are possible effects on course success.

References

1. Adamic LA, Zhang J, Bakshy E, Ackerman MS (2008) Knowledge sharing and yahoo answers: everyone knows something. In: Proceedings of the 17th international conference on World Wide Web. ACM, New York (2008), pp 665–674
2. Anderson A, Huttenlocher D, Kleinberg J, Leskovec J (2014) Engaging with massive online courses. In: Proceedings of the 23rd international conference on World Wide Web. ACM, New York, pp 687–698. doi:10.1145/2566486.2568042
3. Bader BW, Harshman RA, Kolda TG (2007) Temporal analysis of semantic graphs using ASALSAN. In: 7th IEEE international conference on data mining. ICDM 2007, pp 33–42. doi:10.1109/ICDM.2007.54
4. Borgatti SP, Everett MG (1993) Two algorithms for computing regular equivalence. Soc Networks 15(4):361–376. doi:10.1016/0378-8733(93)90012-A
5. Breiman L (2001) Random forests. Mach Learn 45(1):5–32. doi:10.1023/A:1010933404324

6. Clow D (2013) MOOCs and the funnel of participation. In: Proceedings of the 3rd international conference on learning analytics and knowledge, LAK'13, pp 185–189. ACM, New York. doi:10.1145/2460296.2460332
7. Cortes C, Pregibon D, Volinsky C (2001) Communities of interest. In: Proceedings of the 4th international conference on advances in intelligent data analysis (IDA), pp 105–114
8. Cui Y, Wise AF (2015) Identifying content-related threads in MOOC discussion forums. In: Proceedings of the 2nd ACM conference on learning @ scale, L@S'15. ACM, New York, pp 299–303. doi:10.1145/2724660.2728679
9. Doreian P, Batagelj V, Ferligoj A, Granovetter M (2004) Generalized blockmodeling (Structural analysis in the social sciences). Cambridge University Press, New York
10. Dunlavy DM, Kolda TG, Acar E (2011) Temporal link prediction using matrix and tensor factorizations. ACM Trans Knowl Discov Data 5(2):10:1–10:27. doi:10.1145/1921632.1921636
11. Furtado A, Andrade N, Oliveira N, Brasileiro F (2013) Contributor profiles, their dynamics, and their importance in five Q&A sites. In: Proceedings of the 2013 conference on computer supported cooperative work. ACM, New York, pp 1237–1252
12. Gauvin L, Panisson A, Cattuto C (2014) Detecting the community structure and activity patterns of temporal networks: a non-negative tensor factorization approach. PLoS ONE 9(1):e86028. doi:10.1371/journal.pone.0086028
13. Gillani N, Eynon R (2014) Communication patterns in massively open online courses. Internet High Educ 23:18–26. doi:10.1016/j.iheduc.2014.05.004
14. Gillani N, Yasseri T, Eynon R, Hjorth I (2014) Structural limitations of learning in a crowd: communication vulnerability and information diffusion in MOOCs. Sci Rep 4:6447. doi:10.1038/srep06447
15. Harrer A, Schmidt A (2013) Blockmodelling and role analysis in multi-relational networks. Soc Netw Anal Min 3(3):701–719. doi:10.1007/s13278-013-0116-x
16. Harrer A, Zeini S, Ziebarth S (2010) Visualisation of the dynamics for longitudinal analysis of computer-mediated social networks-concept and exemplary cases. In: Memon N, Alhajj R (eds) From sociology to computing in social networks: theory, foundations and applications. Springer, Vienna, pp 119–134. doi:10.1007/978-3-7091-0294-7_7
17. Hecking T, Hoppe HU, Harrer A (2015) Uncovering the structure of knowledge exchange in a MOOC discussion forum. In: Proceedings of the 2015 IEEE/ACM international conference on advances in social networks analysis and mining 2015, ASONAM'15. ACM, New York, pp 1614–1615. doi:10.1145/2808797.2809359
18. Hoppe HU, Göhnert T, Steinert L, Charles C (2014) A web-based tool for communication flow analysis of online chats. CEUR. http://ceur-ws.org/Vol-1137/LAK14CLA_submission_6.pdf
19. Kaufman L, Rousseeuw P (1987) Clustering by means of medoids. In: Dodge Y (ed) Statistical data analysis based on the L1-norm and related methods. North-Holland, Amsterdam, pp 406–416
20. Kim J, Shaw E, Feng D, Beal C, Hovy E (2006) Modeling and assessing student activities in on-line discussions. In: Proceedings of the AAAI workshop on educational data mining
21. Kim SN, Wang L, Baldwin T (2010) Tagging and linking web forum posts. In: Proceedings of the 14th conference on computational natural language learning, CoNLL'10. Association for Computational Linguistics, Stroudsburg, pp 192–202
22. Kolda TG, Bader BW (2009) Tensor decompositions and applications. SIAM Rev 51(3):455–500. doi:10.1137/07070111X
23. Lerner J (2005) Role assignments. In: Brandes U, Erlebach T (eds) Network analysis. Lecture notes in computer science, vol 3418. Springer, Berlin, pp 216–252. doi:10.1007/978-3-540-31955-9_9
24. Mitchell M (1998) An introduction to genetic algorithms. MIT Press, Cambridge
25. Nickel M, Tresp V, Kriegel HP (2011) A three-way model for collective learning on multi-relational data. In: Proceedings of the 28th international conference on machine learning, pp 809–816

26. Ó Duinn P, Bridge D (2014) Collective classification of posts to internet forums. Case-based reasoning research and development, vol 8765. Springer, Cham, pp 330–344. doi:10.1007/978-3-319-11209-1_24

27. Pattison P (2009) Algebraic models for social networks. Encyclopedia of complexity and systems science. Springer, New York, pp 8291–8306. doi:10.1007/978-0-387-30440-3_492

28. Ramesh A, Goldwasser D, Huang B, Daume III, H, Getoor L (2014) Understanding MOOC discussion forums using seeded LDA. In: Proceedings of the 9th ACL workshop on innovative use of NLP for building educational applications. ACL, Portland

29. Rossi LA, Gnawali O (2014) Language independent analysis and classification of discussion threads in Coursera MOOC forums. In: Proceedings of the 15th IEEE international conference on information reuse and integration, pp 654–661. doi:10.1109/IRI.2014.7051952

30. Stump G, DeBoer J, Whittinghill J, Breslow L (2013) Development of a framework to classify MOOC discussion forum posts: methodology and challenges. Report (available online (26/09/2016)). https://tll.mit.edu/sites/default/files/library/Coding_a_MOOC_Discussion_Forum.pdf

31. Wasserman S, Faust K (1994) Social network analysis: methods and applications. Structural analysis in the social sciences, vol 1. Cambridge University Press, Cambridge

32. White DR, Reitz KP (1983) Graph and semigroup homomorphisms on networks of relations. Soc Networks 5(2):193–234

33. Whittaker S, Terveen L, Hill W, Cherny L (2003) The dynamics of mass interaction. In: From Usenet to CoWebs. Springer, London, pp 79–91

34. Wiesberg S, Reinelt G (2015) Relaxations in practical clustering and blockmodeling. Informatica 39(3):249–256

35. Wong JS, Pursel B, Divinsky A, Jansen BJ (2015) An analysis of MOOC discussion forum interactions from the most active users. Social computing, behavioral-cultural modeling, and prediction, vol 9021. Springer, Cham, pp 452–457. doi:10.1007/978-3-319-16268-3_58

36. Zeini S, Göhnert T, Hecking T, Krempel L, Hoppe HU (2014) The impact of measurement time on subgroup detection in online communities. In: Can F, Özyer T, Polat F (eds) State of the art applications of social network analysis. Springer, Cham, pp 249–268. doi:10.1007/978-3-319-05912-9_12

37. Zhang J, Ackerman MS, Adamic L (2007) Expertise networks in online communities: structure and algorithms. In: Proceedings of the 16th international conference on World Wide Web, WWW'07. ACM, New York, pp 221–230. doi:10.1145/1242572.1242603

Chapter 9
Diffusion Process in a Multi-Dimension Networks: Generating, Modelling, and Simulation

Youssef Bouanan, Mathilde Forestier, Judicael Ribault, Gregory Zacharewicz, and Bruno Vallespir

1 Introduction

Nowadays, many researchers are interested in developing new and more efficient systems for social simulation. The issues explored include psychology, organizational behavior, sociology, political science, economics, anthropology, geography, engineering, archaeology, and linguistics [1, 2]. For instance, military simulation systems have investigated to support deeply detailed analysis of individual's behavior. The training simulation systems currently in use by the military forces have been designed for conventional warfare. The aim of the paper is to model a population of interconnected individuals in order to simulate and observe the reaction of this population face to different operations.

Today, most of the marketing, polls, and influence operations and in general information spreading involve heterogonous population structure and targets. The objective is to anticipate by simulation and to observe the impact of information over a population.

So, the aim of the paper is to model a population with cultural features describing and simulating the effects (or influence) of some operations of information diffusion within it. Modelling population needs the understanding of the specific norms and ways the society organizes itself, e.g., an Eastern European society does not have the same features as an Asian, Western Europe, or African one (even inside each society cited subcategory can be defined). The model of the population has to respect the codes of the society in order to accurately simulate the information diffusion. Furthermore, the relationships between people are too complex to be modeled by one link. In general, most real and engineered systems include multiple subsystems and layers of connectivity, and it is important to take such features into

Y. Bouanan (✉) • M. Forestier • J. Ribault • G. Zacharewicz • B. Vallespir
University of Bordeaux, IMS, UMR 5218, 33400, Talence, France
e-mail: youssef.bouanan@ims-bordeaux.fr

© Springer International Publishing AG 2017   199
J. Kawash et al. (eds.), *Prediction and Inference from Social Networks and Social Media*, Lecture Notes in Social Networks, DOI 10.1007/978-3-319-51049-1_9

account when trying to obtain a complete understanding of them. Social networks are generally based on only one relationship between people, or an aggregation of several relationships into one. They are part of networks that we call them multiplex. Multiplex networks are networks in layers and with connections between layers; the interconnections between layers are only between a node and its counterpart in the other layer or better in the same node [3]. Multidimensional social networks (MSN) begin to emerge due to the importance of each relationship in the communication process [4–6].

Flatten an MSN into a one dimension social network does not allow to consider each relation as a unique way to communicate with its own communication rules. It also does not allow representing the complexity of an individual social life. In this paper, an MSN is modeled with the idea that information disseminates differently according to the link through which the information propagates: people do not receive and transmit in the same way information according to the person who gave them the information. Following this postulate, this paper presents the general framework of our MSN in order to generate a model of a population based on several relationships. Then, using social science research, some relationships representing a part of the human social life are described.

In [7], the authors discussed how a MSN is generated in the case of sub-Saharan Africa societies. The focus of this paper is the evolution of work published in [7]. It proposes a general framework to model a population based on its social structure rules and its cultural features and to simulate the propagation phenomenon in networks. In more details, the authors propose algorithms for generating new dimension "political partisan dimension" and "religious partisan dimension" in the case of sub-Saharan Africa. They also add dynamics on the MSN by simulating message dissemination and its impact in individuals who receive the message. This framework is really adaptable thanks to the MSN architecture that separate and distinguish the relationships the one from the others, to add or delete a dimension, and to set for each dimension its own features and diffusion rules. The use of an MSN allows us also to model the message acceptation and the transmission rules for each dimension.

The remainder of this paper proceeds as follows. Section 2 presents the concepts of multilayer network and dissemination processes in multilayer networks. Section 3 introduces the related work in the context of social network and diffusion processes. Section 4 describes our methodology: how nodes are created, what are our dimensions, how people are linked, and so on. Then, Sect. 5 presents the measures of our MSN and our proxy/server architecture of simulation of message diffusion in an MSN. Finally, the paper concludes with directions for future work.

2 Preliminaries

In this section the authors introduce the concepts of multilayer network and dissemination processes in multilayer networks.

A simple network (i.e., a single-layer network) can be represented by a graph [8, 9]. A graph is a tuple $G = (V; E)$, where V is set of nodes (vertices) representing social entities: humans, organizations, departments, etc., called also actors, agents, or instances and $E \subseteq V \times V$ is a set of (ordered or unordered) edges (arcs, connections, or ties) that connects pair of nodes. Since social networks usually represent one kind of relationships, they are also called single-layered social network SSN [10]. A multilayer network is a data structure made of multiple layers, where each layer is a monoplex network. A multi-layered social network MSN is a network extended to multiple edges between pairs of nodes/actors. It is defined as a tuple <V, E, L> where:

- V – is a set of actors (social entities);
- E – is a set of tuples <v,e,l>, l∈L, v≠e;
- L – is a set of distinct layers (types of relationships).

Each layer in the MSN corresponds to one type of relationships between people. Different relationships can result from the character of connections, types of communication channel. The examples of different relationships can be: friendship, family, or work. Different communication channels that result in different types of connections are: email exchange, VoIP calls, instant messenger chats, etc. These networks can be defined as a sequence of graphs:

$$\{G_\gamma\}_\gamma{}^\delta = \{(V_\gamma, E_\gamma)\}_\gamma{}^\delta{}_{=1} \tag{9.1}$$

where $E_\gamma \subseteq V_\gamma \times V_\gamma$ is the set of edges and γ indexes the graphs or the type of relationships. In our networks, the nodes (agents) are the same across the different layers (i.e., $V_\gamma = V_\delta \; \forall \gamma, \delta$).

In Fig. 9.1, the example of three-layered social network is presented. The set of agents are connected with each other on three layers: L1 (family), L2 (friendship), and L3 (work).

Fig. 9.1 An example of the MSN

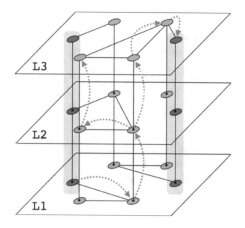

Table 9.1 Different possibilities for spreading an element in a multilayer network

	Same node	Other node
Same layer	Unused	The item spreads to a different node on the same layer
Other layer	The item switches layer but remains on the same node	The element continues spreading in a different node on the another layer

Diffusion processes in networks have a long history in social sciences [11]. With the advent of sufficient storage and computational power, this network diffusion process became an emerging research area in computer science [12]. Propagation models are designed to reproduce phenomena observed in social networks with applications in viral marketing, spread of disease, and diffusion of ideas and innovations. Most models proposed recently are extensions from the independent cascade (IC) [13] and the linear threshold models (LT) [14]. The two models characterize two different aspects of social interaction. The IC model focuses on individual (and independent) interaction and influences among friends in a social network. The IC models can also be identified with the so-called Susceptible/Infective/Recovered (SIR) model for the spread of disease in a network [15]. The LT model focuses on the threshold behavior in influence propagation, which it can be frequently related to; when enough of your friends bought a new phone, played a new computer game, or used new online social networks, you may be converted to follow the same action.

One of the important ideas in the context of dissemination processes in multilayer networks is the fact that elements can also spread from one layer to another [16]. In general there are four possibilities for an element to traverse a multilayer network (Table 9.1): *same-node inter-layer*, when the cascade switches layer but remains on the same node, e.g., when an element is shared on different layers by the same actor; *other-node inter-layer*, when a cascade continues spreading to another node in another layer. In third type, *other-node intra-layer*, the cascade continues spreading through the same layer, e.g., spreading an information between friends. The fourth combination, *same-node intra-layer*, is generally not considered meaningful and therefore omitted in all the diffusion studies that have been considered.

3 Background and Related Work

Social network generation, or more specifically graph generation, appeared in 1960 with the Erdos and Rényi [17] from this time to nowadays, researchers have looked to improve the graph generation to respect social network features [18]: degree distribution, assortativity, clustering coefficient, small world structure [19], and group structure. Social networks are not random graph and have special features to respect in order to generate realistic population. Thus, researchers aim to generate automatically social networks according to these features. Following Erdos and Rényi [17], the dyadic models appeared [20, 21], the Markov random graph [22], and the network generation with small world [23]. Since 2000s, models are improved to take into account the time effect with the notion of preferential attachment: new nodes link proportionally to high connected previous node in a richer get richer configuration [24]. Newman et al. develop algorithms to create social networks with an arbitrary degree distribution [18]. These algorithms allow us to set a distribution degree in order to obtain nodes with a correct number of links. Nowadays, researchers develop methods which could be used for managing several properties as Badham and Stocker who manage the degree distribution, the clustering coefficient, and the assortativity [25].

In a military perspective, Svenmarck et al. simulate the diffusion of information in a fictive country [2]. They model the XL and population using Hofstede's cultural dimension [26], with two religions and several ethnic groups while taking into account the territorial limitations as the access to drinking water. They use a social network where the nodes represent communities. Each community can get over 110 people with an average of 60. After generating nodes and giving them a spatial localization, the authors created links according to their features. The paper addresses two issues to their work: (1) using nodes to represent a community that do not take into account the human process of acceptance or rejection of an information neither than the ability to transmit it to their network; (2) authors use only one relation between nodes without taking into account the different domains of social life and that people share information differently depending on the network.

So, based on this consideration, this work defines an MSN.

Since a few decades, research in sociology described MSN as in [27]. We can find different terminology: it can be fined in literature the terms multilayer networks [3], networks of networks [28], multidimensional networks [29] multiplex networks [30], or multi-relational networks [23], and so on. As this work is based on social networks, using social science theories, the paper will use the closest concept to the proposed work, namely the term MSN. Furthermore the work of Berlingerio et al. gives us strong foundations in MSN architecture [4, 29].

Although social networks have long existed, multidimensional formalization and modelling is fairly recent. Very few works about this subject can be found in literature as stated in [4–6]. In their paper, Pappalardo et al. propose an MSN where relations between people come from three websites of social networking: Foursquare, Twitter, and Facebook [6]. Then, they try to measure the strength of these links. As for Berlingerio et al., they analyze hubs in a multidimensional network [4]. Actually, measures from social network analysis have to be adjusted to MSN. Finally, Forestier et al. propose an MSN from online discussions where relations are from discussion structure and text content. These relations help to find celebrities in the discussions [5].

So, MSN model has a really big interest in catching the social life diversity of different kinds of people.

4 Materials and Methods

The SICOMORES project aims to simulate the actions of influence happening in the context of stabilization phase of modern asymmetric conflicts. These actions of influence can be Psychological Operations (PSYOPS) or Civil Military Cooperation (CIMIC). Both of these operations aim to convince people's heart and mind. These operations aim to make the local population an ally. Military spread messages to influence info-targets in their behavior.

In order to simulate the message diffusion, this work defines an MSN. Berlingerio et al. defined in [29] a multidimensional network as a triple G = (V,E,L) where: V is a set of nodes; L is a set of labels; E is a set of labeled edges, i.e., the set of triples (u,v,d) where u,v ∈ V are nodes and d ∈ L is a label. Also, the paper uses the term dimension to indicate label. Each dimension has its own generation rules according to social science theories. Some rules are defined to generate relationship between people according to their features and their socialization rules. The greatest benefit of using an MSN is to keep distinct all the socialization dimensions. It implies three major improvements:

- Generate relationships according to sociability rules, e.g., families structures are different from friendship structures;
- MSN proposes a very flexible architecture: dimensions can be easily added or deleted;
- Using an MSN allows to define propagation rules for each dimension allowing a more accurate information diffusion simulation.

So, this section presents how to generate nodes and the dimensions linking them.

4.1 Human Terrain

Each node in our system represents an intelligent individual. For an optimal dissemination of information process, two elements are of paramount importance: integrating the cognitive processes of message treatment involving individual features and network configuration and correctly capturing and representing a relevant socio-cultural context, i.e., the complex sociality of each and every person, in the form of a multiplex network.

Each individual in the system is an instance k of the INDIVIDUAL frame and so described by a set of attributes and variables:

- Social features: Gender, age, social level [25, 31], religion, ethnicity, political opinion, role in the family, leader status.
- Cultural features: Cultural values system (type of cultural feature, cultural feature, importance [25, 31]).
- Reachability features: Language(s), literacy, reachable by radio, reachable by television, reachable by text message.

- Psychological features: Opinion toward the force [25, 31], intellectual level [25, 31], needs (need, satisfaction degree [25, 31]).

The role in the family can be husband, wife, child, or related individual and different family configurations are simulated (nuclear, enlarged, extended). The political opinion indicates belonging to one of the political factions represented on the political layer. Some agents are assigned specific roles such as head of family, political leader, or religious leader. A head of family can be feminine or masculine but can only be leader if masculine, and a leader can be either political or religious, but not both at the same time. The porosity and interlinkage between the religious and political domains characterizing Africa today is represented by similar structural properties in the religious and political layers of the network itself.

Overall, the different layers in our network are: family, neighborhood, friendship, partisan (political), partisan (religious), and war-time (representing links generated by communitarianism and violence, religious, or ethnic). To be accurate to actual diffusion process, algorithms ask for certain information such as the proportion of each ethnicity in the group or the proportion of each kind of family (this work has defined in Sect. 2, three categories of family). People can also speak several languages but at the time of writing only one language has been taken into account, i.e., the one the most commonly spoken and understood. For example, if an ethnic group not only possesses its own language or dialect but also speaks English, only the English language is taken into account since it is the intergroup communication means.

At first, the authors thought to assign features to node in a fully random way but it quickly appeared as a nonrealistic idea: individual's features are dependent on each other, e.g., in some cultures, illiteracy can be relative of the social level or the gender. So it has been improved to interdependent data generating according to population information available. Finally, the last static information concerns the access to various communication media. It is necessary to model the individual's accessibility for media such as radio, TV, cell phone, and internet. This information of accessibility helps user to choose the pertinent medium to reach the info-targets. In the process of information diffusion it is really important to consider the medium used, all the media do not have the same impact on the population: diffusing a message during a famous TV shows could have more repercussion than in a radio show [32]. At the end, if the final users do not have the information described previously, the tool generates it randomly.

4.2 The Three Primary Dimensions: Family, Friends, and Neighbors

This work defines three permanent dimensions L = {family, friends, and neighbor} according to [31]. Figure 9.2 shows the three first social relationships. Each individual, represented by a node, is presented on each dimension of the MSN (even

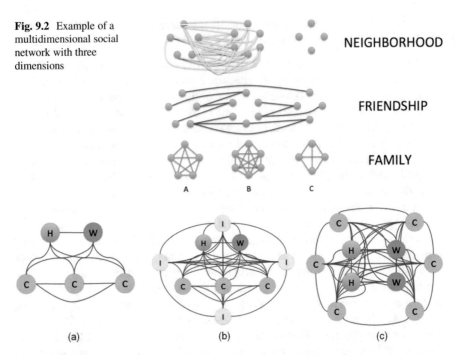

Fig. 9.2 Example of a multidimensional social network with three dimensions

NEIGHBORHOOD

FRIENDSHIP

FAMILY

A   B   C

(a)   (b)   (c)

Fig. 9.3 The three family structures. (**a**) Nuclear family, (**b**) Enlarged family, (**c**) Extended family

if he has any relationship in the dimension). People can be linked in one or more dimensions. Each node is present in each dimension but can be isolated in some dimensions, e.g., a person may have no family, no known neighbors.

There are three distinct families on the first dimension (the family dimension). The second dimension shows how people are linked in the friendship relationship and finally, the way people are linked in the neighborhood dimension. The following three subsections present the generation of each dimension.

- Family dimension: Families are considered as the basis of each society over the world. They can be structured in different ways but this dimension belongs to the three first socialization groups. On the overall possibilities throughout the world three family structures have been selected as shown in Fig. 9.3. As expressed in the figure, more than the size of the families, these are the roles of family members that are important. These roles have a big influence in the message spreading. Figure 9.3a presents a nuclear family composed by a pair of adults (husband H and wife W) and their children (C); Fig. 9.3b presents the enlarged family that can be defined as a nuclear family with some related individuals (I). Finally, Fig. 9.3c presents extended family, i.e., a family composed of several married couples and their children.

Some families can be polygamous. In this case, families are still one of these three structures but with the addition of wives (a husband can have two wives).

During the generation of the family dimension, features, such as a common ethnicity and language, have been assigned to nodes belonging to the same family to respect certain cohesion. Algorithm 1 shows how the families have been build. According to the quantity of each family previously introduced, it generates families with socio-cultural features (e.g., ethnicity, religion, social level).

Algorithm 1: Family Links Generation

```
Algorithm GenerationFamilyDimension(int nbIndividus)
  int perSeules, proportionFamilleNucléaire,
  proportionFamilleEtendue, proportionFamilleEtendue,
  proportionFamilleElargie, proportionMatri, proportionPoly ;
  While (((Number of person ∈ family) - singlePer) > 0)
    do
         String ethnie, langue, religion;
           if(proportion of matriarchal families <
         proportionMatri) then
              bool matri   := true
           else if (proportion of polygamous families <
         proportionMatri) then
              bool poly   := true
           end
         // Constructing the families
           if (proportionOfNuclearFamilies < number ofnuclear
         families) then
         Generation Nuclear Family with (ethnie, religion,
         langue, matri, polygame)
         Else if(proportionOFExtendedFamily < number of ex-
         tended families) then
         Generation Extended Family with (ethnie, religion,
         langue, matri, polygame)
         Else if (proportionOfEnlargedFamily < number of en-
         larged families) then
         Generation Enlarged Family with (ethnie, religion,
         langue, matri, polygame)
         End if
    End while
  // Generating single persons
  for (each lonely node) do
  int sexe, age
  String ethnie, langue, religion ;
  node singlPerson = setAttributs(sexe, age, leaderFamily,
  Id_individu, ethnie, religion, langue)
    End For
End Algorithm
```

For now, it is assumed that a family shares a same religion, a same social level but these assumptions may change in future improvements. For example, in some parts of the world, finding several religions inside a same family is common. Concerning the reachability of the families, it is assumed that if at least one member of the family is reachable by TV or radio, all the other members of the family are also reachable. Finally, some people do not belong to a family. For example, at war-time it can be frequent, especially in refugee camps. Thus, some nodes do not have any family link. They have been called lonely nodes in Fig. 9.4.

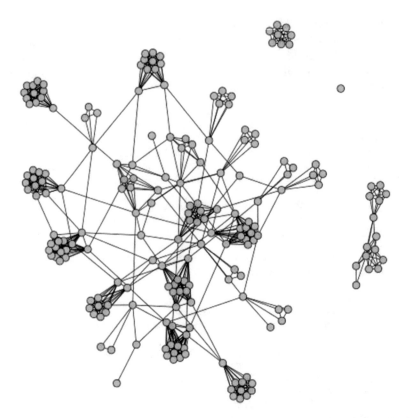

Fig. 9.4 Generation of the family dimension represented by a population of 200 people with 10% of lonely nodes

Figure 9.4 shows a generation of families for a group of 200 people with 10% of lonely nodes, i.e., nodes without family. It clearly shows that each family is a clique and can be nuclear, extended, or enlarged.

- Friendship links: Then the friendship dimension is generated. Friendship is based on homophily following the well-known motto birds of a feather flock together, i.e., the way that people tend to link with similar people [33]. A measure of homophily has been proposed based on the individual's attributes as explained in the following equation:

$$FriendshipHomophily = w_s sexe + w_a age + w_{sc} socialLevel +$$
$$w_e ethnicity + w_l language + w_r religion$$

(9.2)

Each parameter is multiplied by a factor in order to easily advantage or disadvantages one, e.g., in some parts of the world, ethnicity can be more important to become friends than religion or vice versa.

Algorithm 2 presents how friendship links are created in its respective dimension. It first creates link between randomly chosen nodes if their homophily score is higher than a given threshold (homophilyThreshold). Then homophily is measured according to the following Eq. (9.2). This operation is repeated until to reach the half of the average number of friends (AverageFriendsThresold). Then clustering coefficient is raised because friendship dimension has to have a higher coefficient clustering: friends of my friends are my friends.

Algorithm 2: Friendship Links Generation

```
Algorithm GenerationFriendshipDimension(int AverageOfFriends)
   Data: Graph frienGraph; int AverageFriendsThreshold;
   double homophilyThreshold; double ccThreshold;
   While (AverageOfFriends < AverageFriendsThreshold) do
        node1 <- random node of friendGraph;
        node2 <- random node of friendGraph;
        if((homophily(node1,node2) / maxHomophily) >
   homophilyThreshold) then
            Create FriendshipLink(node1, node2);
        End if
     End while
   // IMPROVE COEFFICIENT CLUSTERING
   While (AverageOfFriends < AverageFriendsThreshold) do
        randomNode<- random node of friendGraph;
        egoFrienGraph <- ego-network of randomNode in the
   friendGraph;
        while (CoefficientClustering(egoFriendGraph) <
   ccThreshold) do
            node1 <- random node of egoFriendGraph;
            node2 <- random node of egoFriendGraph;
            Create FriendshipLink(node1, node2);
        End while
     End while
End Algorithm
```

To make this happen, links have added between nodes of node's ego-network. For each randomly chosen node, the ego-network in the friendship dimension is obtained and links are added to obtain a clustering coefficient of ccThreshold on this subgraph. The addition of links is repeated until to reach the average number of friends previously computed.

Figure 9.5 presents the friendship dimension before and after the improvement of the clustering coefficient. It can be seen in Fig. 9.5b that new links exist between unconnected people in Fig. 9.5a.

- *Neighbor links:* The notion of neighbor can be really difficult to apprehend even in one specific culture. People can interpret neighborhood as considering either only people living in the same building level, or in the same building, or in an area of several buildings, or in perimeter of several kilometers. In the approach, the friendship dimension is built on an average number of "social homophily" neighbors (as shown in Fig. 9.5). For that purpose, four attributes are selected

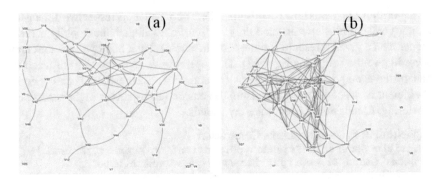

Fig. 9.5 The friendship dimensions; (**a**) before the improvement of clustering coefficient; (**b**) after the improvement of clustering coefficient

to model a neighborhood: the social level, the language, the ethnicity, and the religion. The equation of homophily is created between neighbors as follows:

$$NeighborHomophily = w_{sc}socialLevel + w_{e}ethnicity + w_{l}language + w_{r}religion$$
(9.3)

As for the friendship, the parameters are weighted. For example, we can find some rich and poor neighbors in Europe while the ethnicity could be more important in Africa. The possibility to choose the importance of each parameter allows a fine representation of a population.

Algorithm 3 describes how neighborhood links have been created. First it randomly picks two nodes in the neighborhood graph. If the homophily of these two nodes is higher than a certain threshold (homophilyThreshold), it generates a neighborhood link not only between them but also between their respective families. As a family is defined as a located group of people, each member of the first node's family is necessarily a neighbor of the second node's family. It repeats this operation until to reach the average number of neighbors given in input. Note that neighborhood links are created firstly for the lonely nodes: a person without family should firstly try to socialize with the closest located people.

Algorithm 3: Neighborhood Links Generation

```
Algorithm GenerationNeighborhoodDimension(int distance)
    Data: Graph neighborhoodGraph; int AverageNeighborThreshold;
    double homophilyThreshold;
    While (AverageOfneighbors < AverageneighborsThreshold)
    do
        node1 <- random node of neighborhoodGraph;
        node2 <- random node of neighborhoodGraph;
        if((neighborHomophily(node1,node2) / maxHomophily)
    > homophilyThreshold) then
            Create neighborhoodLink(node1, node2);
            create neighborhoodLink(node1's family, ; node2's
    family);
        end if
    End while
End Algorithm
```

4.3   The Partisan Association Dimensions

Partisan associations represent the networks in which people are involved because of their belief. Two separate partisan associations have been defined: the religious and the political partisan association. These two dimensions have a similar shape because their structures are similar: partisan associations represent groups of people with a leader (political or religious).

- The political partisan association dimension represents groups of people following the idea of a same leader. So the algorithm to generate a political partisan association is the following:

Algorithm 4: Political partisan association generation
```
Algorithm GeneratePoliticalPartisanAssociation ()
   While (each individual does not belong to a group) Do
      Alea   := random number between 20 et 80 //size of the group
      For (i   := 1 to Alea) Do
        Group:= add a node to the group
      End For
      Generate random links between members of the group
// Look for the most connected node in the group
   Leader   := first node of the group
      For (each node of the group) Do
         If (node's degree > leader's degree) Then
            Leader   := node
         End if
      End For
      Connect the leader to all the members of the group
   End While
End Algorithm
```

The previous algorithm has the aim to create groups of people inside the whole population. But, with the algorithm, the groups are totally disconnected from each other. This not reflects the reality for political groups where groups are connected. So, the groups are connected via the leaders (see Fig. 9.6) using the following algorithm:

```
Algorithm GenerateLinkBetweenLeaders()
      Get the leader of each group
      While (average numbers between two leaders < 4) Do
        Generate link between leaders
      End While
End Algorithm
```

Figure 9.6 shows the political partisan association with groups linked with their leader.

- The religious partisan association dimension can represent a church, a mosque, a sectarian movement, and so on. The idea behind this dimension is that people exchange information during religious services, for example. So, this dimension has a real influence in the message diffusion.

Fig. 9.6 Political partisan
association

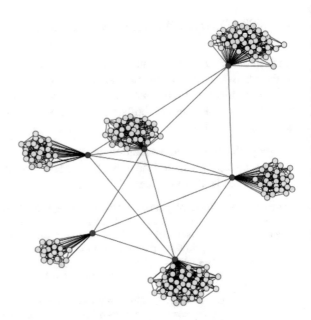

Algorithm 5: Religion Partisan Association Generation

```
Algorithm GenerateReligiousPartisanAssociation ()
    For each religion in the population Do
       Get all the individuals sharing this religion
       While (all individuals do not belong to a group) Do
          Alea  := random number between 20 and 80
          For (i    := 1 to Alea) Do
             Group   := add a node to the Group
          End For
Generate random links between people in the group
          Leader    := First node of the group
          For (Each node in the group) Do
             If (nodes's degree > leader's degree) Then
                Leader   := node
             End if
          End For
          Connect the leader to each member of the group
       End while
    End For
End algorithm
```

Algorithm 5 presents how to generate the religious groups. But, as for the political partisan association, the group cannot be totally disconnected. So, the same algorithm as the political partisan association is used just adding one condition: to be linked together, leaders have to share the same religion.

Figure 9.7 shows the result of the religious partisan association. The color of the nodes represents the religion (pink and blue). As displayed, the groups with pink color can only be connected to pink groups and orange groups can only be connected to orange one.

Fig. 9.7 Religious partisan
association

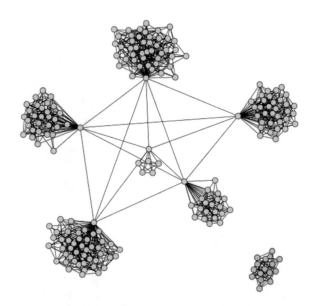

4.4 The War Time Dimensions

The previous sections have just presented the three primary dimensions of our MSN. These three dimensions form the basis of social life and should be represented for all the future population to generate. However, new forms of socialization could appear according to the situation of an area studied, for instance, in freshly stabilized zone after a war. So this study defines two kinds of socialization: the ethnicity and the religious. When the situation is critical, people tend to segregate according to a kind of conflict.

For now, two forms of communitarianism have been modeled: the religion one and the ethnic one. During this critical phase, the dimension can be activated according to a chosen context. So this new socialization could dominate the three primary kinds of socialization (even if they still exist). One of the best assets of MSN is the possibility to easily add or delete dimensions to better represent interaction between people.

Algorithm 6 presents how to generate the war-time dimension. It first creates the link in the war-time dimension using previously created links between nodes and their friends and neighbors who share the same ethnicity or religion (according to the segregation choice). Then it raises the coefficient clustering by adding links between nodes sharing a same ego-network. In this dimension, it does not exit a way to communicate between two nodes with a different religion or ethnicity.

Algorithm 6: The war-time links generation

```
Algorithm GenerationWarTimeDimension(int AverageOfFriends)
Data: Graph warTimeGraph; int AverageThreshold;double ccThreshold,
segregationChoice;
```

Fig. 9.8 The war-time
dimension based on the
religion

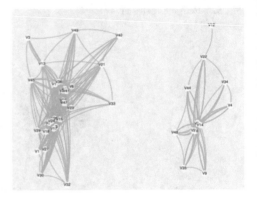

```
for each node in warTimeGraph do
   Retrieve friends and neighbors who share the same
religion or ethnicity;
   Generate Links between each node and its friends and neighbors
who share the same religion or ethnicity; End For
// IMPROVE COEFFICIENT CLUSTERING
While (AverageOfFriends < AverageThreshold) do
   randomNode <- random node of warTimeGraph;
   egoWarTimeGraph <- ego-network of randomNode
   in the warTimeGraph;
   while (CoefficientClustering(egowarTimeGraph) < ccThreshold) do
        node1 <- random node of egoWarTimeGraph;
        node2 <- random node of egoWarTimeGraph;
        Create warTimeLink(node1, node2);
   End while
   End while
End Algorithm
```

Figure 9.8 shows the war-time dimensions based on the religion. It clearly points that people are split into two groups without gate between these two.

4.5 Formalization of Human Behavior

4.5.1 DEVS Formalism

Zeigler introduced the DEVS formalism in the early 1970s for modelling discrete-event systems in a hierarchical and modular way [34]. With DEVS, a model of a large system can be decomposed into smaller component models with coupling specification between them. DEVS formalism defines two kinds of models: (1) atomic models that represent the basic models providing specifications for the dynamics of a subsystem using transition functions (external transition, internal transition); (2) coupled models that describe how to couple several component models together to form a new model. DEVS provides an automatic simulation based on time synchronization and message propagation.

An atomic model allows specifying the behavior of a basic element of a given system. Atomic model behavior is determined by transition functions (internal, external, and confluent), output function, and time advance function:

- Internal transition function (δ_{int}): It describes the autonomous behavior of the model and determines how the states evolve when there is no input. It creates internal events.
- External transition function (δ_{ext}): It describes how the model responds to input and how it will change its state. It creates external events.
- Confluent transition function (δ_{conf}): It handles the simultaneous occurrence of an internal and external event.
- Output function (λ): It gives the output messages of the model that are caused by an internal event.
- Time advance function (ta): It schedules autonomous changes of state; i.e., the next internal event. It determines the maximum lifetime in the current state.

Coupled models meanwhile describe the system's structure. They allow hierarchical level of view and describe the link between models. The communication between models is possible through message exchanges that represent events.

4.5.2 Specification of Message Processing by the Receiver

In the agent-based model, individuals are represented as agents. Each agent is described by a set of attributes distinguished into two categories:

- Static attributes: gender, social status, religion, age class, ethnicity, leadership, and language.
- Dynamic attributes (variables): opinion, interest, un/satisfied needs.

Static attributes are intrinsic or unchanged parameters, i.e., time has no effect on them. Dynamic attributes evolve with time or events. For example, individuals can be reached or not by the information depending on its opinion and the social network configuration.

The model of message processing implemented in our framework based on the persuasion model. To represent the range of processing activity available to message receivers, Petty and Caccioppo [35] introduced the concept of an elaboration likelihood model (ELM). Through a variety of situational and personal characteristics may affect elaboration likelihood, two factors, motivation and ability, have received the most attention from ELM researchers [36, 37]. In our framework, we first compute the degree of similarity between the receiver and the sender of the message then we measure the interest in the message theme expressed by the receiver.

DEVS and its extensions have been chosen to describe the human behavior. DEVS is a well-defined formalism which has numerous advantages (including formal aspect, modular, explicit state, time, etc.) over other formalism in the

modelling of complex dynamic system. The DEVS notation has permitted to model
the individual agents and their interconnections within the different networks.

$$M_{AM} = \{X, Y, S, ta, \delta_{ext}, \delta_{con}, \delta_{int}, \gamma\} \tag{9.4}$$

Figure 9.9 describes the message influence on the individual behavior and
potentially its dissemination using the DEVS specifications. The first phase is used
to configure and initialize the agent' attributes. Then, when the agent is in the
"IDLE" phase, and if it receives an external event from another agent on port
"In_1" (In? Packet), it moves to "phase_0". Here, the agent computes the Degree
of Trust between DT depending on the social pressure, similarity between it and
the sender (based on the religion, language, and age class) and its sensitivity to
the message theme. If the DT is below a certain threshold (to be determined by
experimentation), the receiver does not feel concerned enough to further process the
message and rejects it, it moves to "Idle." In the other case, it moves to "Phase_1".
Then, if the message is still strong enough, the receiver moves to "phase_2" else it
returns to "IDLE." In this phase, the receiver computes its ability to process the
message (APM). If APM is below a certain threshold (to be determined during
experimentation), the receiver will follow a purely peripheral route to persuasion,
it moves to "Phase_3". If not, it is engaging a mainly central processing of the
message, it moves to "Phase_4". This message creates an impact on the individual,
and eventually its behavior depending on the agent's opinion and the relationship
between him and the sender. After a period, processing time, the receiver transmits

```
X = {"In_1"}
Y = {"Out_1"}
S = { 'Init', 'Idle', 'Phase_0', 'Phase_1', 'Phase_2',
   'Phase_3', 'Phase_4'}
ta: 'Init' → 0
   'Idle' → ∞
   'Phase_0' → 0
   'Phase_1' → 0
   'Phase_2' → 0
   'Phase_3' → 0
   'Phase_4' → 0
δₑₓₜ: ('Idle', "In_1") →'Phase_1'
δ_con: δ_con(S,φ) = δ_int(S)
δ_int:'Init '→ 'Idle'
   'Phase_0' → 'Phase_1'
   'Phase_1' → 'Phase_2'
   'Phase_2'→ 'Phase_3'
   'Phase_3' → 'Phase_4'
   'Phase_4' → 'Idle'
γ: 'Phase_4' → "Out_1"
```

Fig. 9.9 DEVS specification for cognitive processing of a message by the receiver

the message on its ego-network. After the contact between receiver and sender, the receiver's variables (opinion, interest, satisfaction) change according to the message content, the sender, and cultural factors.

5 Experiment and Results

5.1 Social Network Measures

The following experiment is based on sociological studies of sub-Saharan African societies. These societies have been selected to highlight some cultural features to be generated inside the populations. For example, the notion of neighborhood is truly important in sub-Saharan social life (while it can have a very poor diffusion impact in some European Area) [38].

For all populations, the three family structures have been generated; an average number of 10 friends, an average of 20 neighbors, and an average of 30 religious connections in the war-time dimension have been chosen.

Table 9.2 presents the average degree in the MSN and the average of shortest path between two nodes in the MSN and in each dimension. Even if the average number of neighbors is very close for the three populations, the average shortest path in the MSN is close to 3 in the population 3 while it is close to 2 for the population 1. Concerning the family dimension, the average shortest path is 1: people inside a family are connected to each other member of the family. The friendship dimension (1) and (2) present the average shortest path (1) before and (2) after increasing the clustering coefficient. The shortest path has been decreased of approximately 1 (a bit less for the population 3). Regarding the religious dimension, the average shortest path does not decrease in the same proportion. It is due to the fact that the population is split into two distinct groups. Finally, concerning the neighbor dimension, the average shortest path increases with the size of the population. Hence the population is close to the small world structure for a population of 5000 individuals.

Table 9.2 Our three generated population

Number of population	Population 1 (500)	Population 2 (1000)	Population 3 (5000)
Average of degree	64	64	69
Min	9	5	4
Max	140	168	190
Average of shortest path between two nodes			
MSN	2.09	2.33	2.76
Family dimension	1	1	1
Friendship dimension (1)	4	4.47	4.59
Friendship dimension (2)	3.08	3.83	3.54
Neighbor dimension	3.4	4.6	5.17
Religious dimension (1)	2.3	2.6	2.95
Religious dimension (1)	2	2.27	2.78

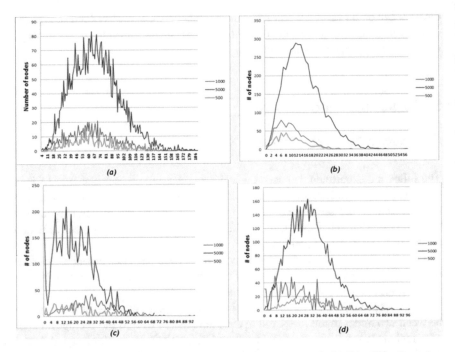

Fig. 9.10 Distribution of degree. (**a**) Multidimensional social network, (**b**) Friendship dimension, (**c**) Neighbor dimension, (**d**) Religion dimension

Figure 9.10 shows the degree distributions for the MSN and for each dimension. Green curves represent the population 1, the blue curves represent the population 2, and the population 3 is in red. The curves look more like a normal law distribution than a power law. This study still has to adjust the algorithms to respect a pure power law distribution. As a perspective in the population generation, less randomizing will be investigated for the choice of the nodes to be linked. The authors also think to avoid lonely nodes in the friendship dimension.

5.2 Using MSN in Simulation

The use of an MSN is fairly new in simulation. This study is using the DEVS formalism implemented with Virtual Laboratory Environment (VLE) [39]. VLE is a software and an API which supports multi-modelling, simulation, and analysis. It addresses the reliability issue by using recent developments in the theory of modelling and simulation proposed by Zeigler. An MSN implementation architecture, extending VLE, and a set of VLE models to simulate the dissemination of information have been previously proposed in [40]. This architecture is based on proxy/server architecture. In more details, this architecture aims to isolate a node (called server-node) to its rules of diffusion (proxies) for each dimension. In this way the server-node contains all the information about the individual such as his opinion, his age, and his religion. It has as many proxies as there are dimensions.

This architecture is flexible and sensitive to a changing environment. As it is easy to add or delete a dimension using an MSN model, it is still easy to add or delete dimensions in our proxy/server architecture without modifying the whole structure. Furthermore, using this proxy/server architecture, each propagation rules can be adapted to each dimension, with the inner idea that information is not transmitted and received in the same way with families, friends, and so on.

5.2.1 Experiment

The MSN concept introduced in this paper is illustrated by modelling and simulating the spread of information in an MSN. The goal of this experiment is to study the diffusion phenomena in the framework of multidimensional networks. The study of the information influence processes initiated by starting point over network member attributes is done regrading dynamically. In more details, a population of 200 agents has been generated and connected on three layers: family, friends, and neighbors. These three layers represent the primary groups defined in Sect. 4.2. Assuming that all agents are starting in a state "0" (they don't received the message yet), and when an agent receives the message, its state changes to "1" if it has an interest to the message or changes to "2" if it receives the message but it has no interest to the message or the item. The agents in the state "2" block the spreading process and they do not update their behavior. However, agents in state "1" update their behavior and their parameters. In this case, the update of agent's opinion is based on the concept of opinion change described by [41]. Social influence occurs when a message emitter and a message receiver get in contact. The message emitter attempts to communicate position about an opinion to the receiver. As result of this conversation, the receiver of the message may shift her/his opinion some distance towards or away from that of the emitter's opinion. This opinion is characterized by the opinion number and the opinion confidence bounds. The study proposed to distinguish two categories of population: extremist and moderate people. The first category is located at the ends of the distribution of opinion (their uncertainty is lower, they block the messages).

Implementation

The implementation task in the cycle of modelling and simulation is the writing of a source code starting from DEVS formal specifications. In VLE, the implementation of a DEVS model is achieved by an inheritance of the DEVS atomic model class, a DS-DEVS executive model class, or another DEVS extension of the VLE framework. We present here the implementation of the models previously defined.

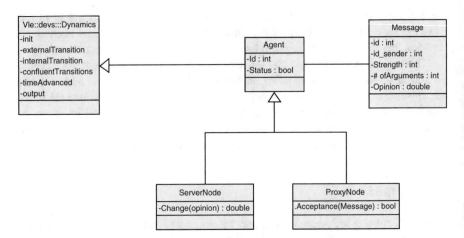

Fig. 9.11 Inheritance tree of the VLE DEVS atomic model devs::Dynamics and the abstract Agent class

The agent is a generic class. ServerNode and ProxyNdore classes inherit from the agent class for the displacement behavior (see Fig. 9.11), and add specificities for the management of node. The agent class provides an Agent::Change function which sends a new opinion to node and an Agent::Acceptance function to check if the agent accepts the message or reject it depending on its attributes, message characteristics, and network configuration.

Simulation

The specification of the experimental frame is given by the VPZ file. It describes all the setup of the system: the graph of coupled models, how models are initialized and how to observe them. In the following example, we use GraphLoader which allows the automatic generation of lattice models (where node is an atomic or coupled DEVS model and connections depending on the type of the relationship between the nodes). Figure 9.12 shows the graphical file done by the GVLE tool to describe our case (message dissemination in an MSN) (Fig. 9.12).

According to the seminal theory of McGuire [32], a message has to catch the motivation of the receiver in order to be processed, with high cognitive activity or not. This step is integrated into our system as a computation of the Degree of Interest (DI). We first compute the degree of similarity between the receiver and the sender. Then, we compute the social pressure measured from the network configuration.

At this level, statistical test is helpful for the validation of this work. Nevertheless even a simple analysis of the result reveals by simulation the possibility to maximize the effect of the message over a population. It can be done by giving the right strength and content value to the message in order to optimize the effects over the population and not to spend too much effort regarding a limited result (Fig. 9.13).

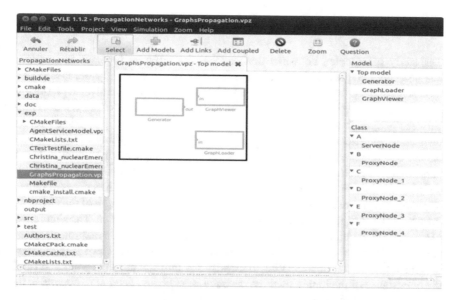

Fig. 9.12 Graphical VPZ file to specify the dissemination of a message in an MSN

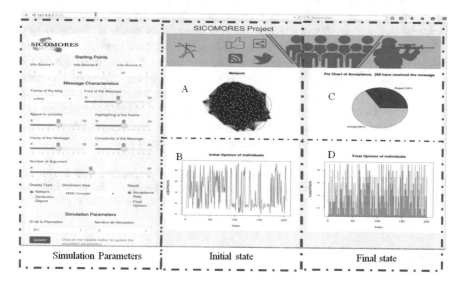

Fig. 9.13 Screenshot of the user interface of our application done by Rstudio using shiny. (**a**) represents the graph of 200 people generated using the algorithms defined before; (**b**) illustrates the initial opinion of 200 individuals; (**c**) illustrates the result of the simulation. It shows the acceptance rate of the agents who received the message and (**d**) is a plot of the final opinion of the individuals

6   Conclusion and Perspectives

The study of complex networks that account for different types of interactions has become a subject of interest in the last few years, especially because of it's representational power in the description of user's interactions in diverse platforms. In this paper, we present the architecture of an MSN for modelling relationships between people. It uses an MSN architecture to generate each dimension with its own rules: the friendship dimension is different than the neighbor dimension and cannot be generated in the same way. So, using an MSN has permitted a better partitioning of each social dimension of human life. Furthermore MSN also allows defining distinct propagation rules for each dimension. The MSN environment implemented proposes features to easily modify such as the strength of the message, the time of the day, the probability of acceptance for each dimension. The simulation has permitted to obtain first result to observe the impact of the population structure and behavior in the message diffusion process. The simulation of two scenarios has presented the interest of simulation to identify key info source in the population and the importance of the tuning over the message strength and content to maximize its impact on the population.

As perspective, the authors will modify the algorithms proposed to obtain distributions of degree following a power law distribution. The first idea is to decrease the randomness in the way that the nodes are chosen. Authors are also thinking about adding links for nodes with small degrees. One of the greatest things using an MSN is its adaptability. The tool makes it easier to delete or add a dimension in respect of the population's features. Finally, the work presented in this paper is part of a military project but opens many possibilities for various applications. Generating a population with an unseen level of cultural features can be used in marketing to simulate the adoption of a new product, in politics to see the diffusion of an idea or the way a politician's reputation changes.

Acknowledgments This work has been partially supported by the SICOMORES Project No. 132936073 funded by French DGA (Direction Générale de l'Armement). It involves the following partners: IMS University of Bordeaux, LSIS University of Marseille, and MASA Group.

References

1. Bouanan Y, El Alaoui MB, Zacharewicz G, Vallespir B (2014) Using DEVS and cell-DEVS for modelling of information impact on individuals in social network. In: IFIP international conference on advances in production management systems. Springer, Berlin, Heidelberg, September 2014, pp 409–416
2. Svenmarck P, Huibregtse JN, van Vliet AJ, van Hemert DA, van Amerongen PJ, Lundin M, Sjoberg E (2010) Message dissemination in social networks for support of information operations planning. TNO defence security and safety, Soesterberg, Netherlands
3. Kivelä M, Arenas A, Barthelemy M, Gleeson JP, Moreno Y, Porter MA (2014) Multilayer networks. J Commun Networks 2(3):203–271

4. Berlingerio M, Coscia M, Giannotti F, Monreale A, Pedreschi D (2011) The pursuit of hubbiness: analysis of hubs in large multidimensional networks. J Comput Sci 2(3):223–237
5. Forestier M, Velcin J, Zighed D (2011) Extracting social networks to understand interaction. In: 2011 international conference on advances in social networks analysis and mining (ASONAM). IEEE, July 2011, pp 213–219
6. Pappalardo L, Rossetti G, Pedreschi D (2012) How well do we know each other? Detecting tie strength in multidimensional social networks. In: IEEE/ACM international conference on advances in social networks analysis and mining (ASONAM). IEEE, August 2012, pp 1040–1045
7. Forestier M, Bergier JY, Bouanan Y, Ribault J, Vallespir B, Faucher C (2015) Generating multidimensional social network to simulate the propagation of information. In: Proceedings of the 2015 IEEE/ACM international conference on advances in social networks analysis and mining. ACM, August 2015, pp 1324–1331
8. Bollobás B (1998) Modern graph theory. Springer, Berlin
9. Newman M (2010) Networks: an introduction. Oxford University Press, Oxford
10. Magnani M, Rossi L (2011) The ML-model for multi-layer social networks. In: 2011 international conference on advances in social networks analysis and mining (ASONAM). IEEE, July 2011, pp 5–12
11. Rogers EM (1962) Diffusion of innovativeness. The Free Press of Glencoe, New York
12. Domingos P (2005) Mining social networks for viral marketing. IEEE Intell Syst 20(1):80–82
13. Goldenberg J, Libai B, Muller E (2001) Talk of the network: a complex systems look at the underlying process of word-of-mouth. Mark Lett 12(3):211–223
14. Granovetter M (1978) Threshold models of collective behavior. Am J Sociol 83:1420–1443
15. Bailey NT (1975) The mathematical theory of infectious diseases and its applications. Charles Griffin & Company Ltd., High Wycombe
16. Min B, Goh KI (2013) Layer-crossing overhead and information spreading in multiplex social networks. preprint arXiv:1307.2967
17. Erdös P, Rényi A (1960) On the evolution of random graphs. Publ Math Inst Hungar Acad Sci 5:17–61
18. Newman ME, Strogatz SH, Watts DJ (2001) Random graphs with arbitrary degree distributions and their applications. Phys Rev E 64(2):026118
19. Milgram S (1967) The small world problem. Psychol Today 2(1):60–67
20. Holland PW, Leinhardt S (1981) An exponential family of probability distributions for directed graphs. J Am Stat Assoc 76(373):33–50
21. Lazega E, Van Duijn M (1997) Position in formal structure, personal characteristics and choices of advisors in a law firm: a logistic regression model for dyadic network data. Soc Networks 19(4):375–397
22. Frank O, Strauss D (1986) Markov graphs. J Am Stat Assoc 81(395):832–842
23. Wasserman S, Faust K (1994) Social network analysis: Methods and applications, vol 8. Cambridge University Press, Cambridge
24. Barabási AL, Albert R (1999) Emergence of scaling in random networks. Science 286(5439):509–512
25. Badham J, Stocker R (2010) A spatial approach to network generation for three properties: degree distribution, clustering coefficient and degree assortativity. J Artif Soc Soc Simul 13(1):11
26. Hofstede G, Hofstede GJ, Minkov M (1991) Cultures and organizations: software of the mind, vol 2. McGraw-Hill, London
27. Krackhardt D (1987) Cognitive social structures. Soc Networks 9(2):109–134
28. D'agostino G, Scala A (2014) Networks of networks: the last frontier of complexity, vol 340. Springer, Cham
29. Berlingerio M, Coscia M, Giannotti F, Monreale A, Pedreschi D (2013) Multidimensional networks: foundations of structural analysis. World Wide Web 16(5–6):567–593
30. Verbrugge LM (1979) Multiplexity in adult friendships. Soc Forces 57(4):1286–1309

31. Cooley CH (1956) Social organization. Transaction Publishers, New Brunswick (U.S.A.) and London (U.K.)
32. McGuire WJ (1969) The nature of attitudes and attitude change. In: The handbook of social psychology, vol 3(2). Addison-Wesley, Reading, MA, pp 136–314
33. McPherson M, Smith-Lovin L, Cook JM (2001) Birds of a feather: homophily in social networks. Annu Rev Sociol 27:415–444
34. Zeigler BP, Praehofer H, Kim TG (2000) Theory of modeling and simulation: integrating discrete event and continuous complex dynamic systems. Academic, San Diego
35. Petty RE, Cacioppo JT (1986) The elaboration likelihood model of persuasion. In: Communication and persuasion. Springer, New York, pp 1–24
36. Bergier J-Y, Faucher C (2016) Persuasive communication from a military force to local civilians: a PsyOps system based on the Elaboration Likelihood Model. In: 15th IEEE international conference on cognitive informatics and cognitive computing. Stanford University, Stanford
37. Stiff JB, Mongeau PA (2003) Persuasive communication. Guilford Press, New York
38. Walther O (2004) Au-delà de l'opposition entre villes et campagnes. Éléments pour un modèle territorial dynamique en Afrique de l'Ouest. L'inform Geogr 68(4):308–319
39. Quesnel G, Duboz R, Ramat É (2009) The virtual laboratory environment—an operational framework for multi-modelling, simulation and analysis of complex dynamical systems. Simul Model Pract Theory 17(4):641–653
40. Bouanan Y, Forestier M, Ribault J, Zacharewicz G, Vallespir B, Moalla N (2015) Simulating information diffusion in a multidimensional social network using the DEVS formalism (WIP). In: Proceedings of the symposium on theory of modeling & simulation: DEVS integrative M&S symposium. Society for Computer Simulation International, April 2015, pp 63–68
41. Friedkin NE, Johnsen EC (1999) Social influence networks and opinion change. Adv Group Process 16(1):1–29

Youssef Bouanan is a Postdoctoral researcher at University of Bordeaux. He received his Ph.D. degree in Production of Engineering from University of Bordeaux, France. His research interests include modeling and simulation in industrial sector, social network and workflow. His email address is bouananyoussef@gmail.com.

Mathilde Forestier is Postdoctoral fellow at the IMS laboratory, University of Bordeaux, France. She obtained her Ph.D. in 2012 from the University of Lyon, France. Her email address is mathilde.forestier@ims-bordeaux.fr.

Judicael Ribault is Postdoctoral fellow at the University of Bordeaux (IUT MP) Lab. IMS. He received his Ph.D. at INRIA Sophia Antipolis, France. He has published several papers in Conferences and is frequent Reviewer in Conferences (SpringSim, TMS/DEVS, etc.). He is involved in European projects. His email address is judicael.ribault@ims-bordeaux.fr.

Gregory Zacharewicz is Associate Professor HDR at University of Bordeaux (IUT MP) with both competences in enterprise engineering and computer sciences. He is recently focused on Enterprise Modelling and Semantic Interoperability. He has published more than 60 papers in international journals and conferences. His email address is gregory.zacharewicz@ims-bordeaux.fr.

Bruno Vallespir is full professor at University of Bordeaux, IMS laboratory. He is member of several international working groups (IFIP, IFAC), he participated to five European projects, has directed more than 20 Ph.D. students and written more than 120 papers in journals and conferences. His email address is bruno.vallespir@ims-bordeaux.fr.

Printed in the United States
By Bookmasters